Praise for *Dambusters*

'A vivid and moving account of the personality clashes, hopes, fears and regrets surrounding one of the most famous bombing operations of all time' *Daily Mail*

'May well rival his seminal *Forgotten Voices of The Great War* . . . His précis of the complex story of how the scientist Barnes Wallis overcame all the obstacles to breaching the Dams, in which British bureaucracy proved as daunting as German efficiency, is a masterpiece of concise storytelling' *Sunday Express*

'A gripping tale capturing the exhilaration of the expedition, while contrasting the sense of loss of 56 men of Bomber Command. A thrilling read for anyone with a nose for a good true tale' *News of the World*

'What a story. And I do not believe that it has ever been better told' Stephen Fry

MAX ARTHUR, who served with the RAF, is the author of *The Sunday Times* bestselling and award-winning *Forgotten Voices of the Great War* and *Forgotten Voices of the Second World War*. His other titles include the classic work on the Falklands campaign, *Above All, Courage*, and *Symbol of Courage: A History of the Victoria Cross*. He is the military obituary writer for the *Independent*.

By the Same Author

Above All Courage: First-hand Accounts from the Falklands Front Line

The Busby Babes: Men of Magic

Faces of World War I

Forgotten Voices of the Great War

Forgotten Voices of the Second World War

Last Post: The Final Word from our First World War Soldiers

Lost Voices of the Edwardians

Men of the Red Beret: Airborne Forces 1940–1990

Northern Ireland: Soldiers Talking

Symbol of Courage: The History of the Victoria Cross

There Shall be Wings – The RAF: 1918 to the Present
(now published as Lost Voices of the Royal Air Force)

The True Glory: The Royal Navy, 1914–1939
and
The Navy: 1939 to the Present Day
(now combined and published as Lost Voices of the Royal Navy)

When This Bloody War Is Over: Soldiers' Songs of the First World War

Dambusters

A Landmark Oral History

Max Arthur

Foreword by Stephen Fry

Published by Virgin Books 2009
2 4 6 8 10 9 7 5 3 1

First published in Great Britain in 2008 by
Virgin Books
Random House, 20 Vauxhall Bridge Road,
London SW1V 2SA

www.virginbooks.com
www.rbooks.co.uk

Addresses for companies within The Random House Group Limited can be found at:
www.randomhouse.co.uk/offices.htm

The Random House Group Limited Reg. No. 954009

A CIP catalogue record for this book
is available from the British Library

ISBN 9780753515730

The Random House Group Limited supports The Forest Stewardship Council [FSC], the leading international forest certification organisation. All our titles that are printed on Greenpeace-approved FSC-certified paper carry the FSC logo.
Our paper procurement policy can be found at www.rbooks.co.uk/environment

Typeset by Palimpsest Book Production Limited, Grangemouth, Stirlingshire
Printed and bound in Great Britain by CPI Bookmarque, Croydon CR0 4TD

This book is dedicated to the 55,573 airmen and ninety-one women of the WAAF who lost their lives while serving with Bomber Command during the Second World War, and in particular, to the fifty-three air crew of 617 Squadron who failed to return from the raid on the dams on the night of 16 to 17 May, 1943.

In recognition of the achievement of Bomber Command, Winston Churchill wrote to Sir Arthur Harris, Air Officer Commander-in-Chief:

'All your operations were planned with great care and skill. They were executed in the face of desperate opposition and appalling hazards, they made a decisive contribution to Germany's final defeat.

'The conduct of the operations demonstrated the fiery gallant spirit which animated your aircrews, and the high sense of duty of all ranks under your command. I believe that the massive achievements of Bomber Command will long be remembered as an example of duty nobly done.'

Author's note

In the writing of *Dambusters* I have listened to many hours of recorded interviews and read a number of personal accounts, and been in contact with the five remaining members of 617 Squadron who flew on the dams raid, as well as ground crew. Throughout the book I have given some historical background, but the heart of the book lies in the personal accounts of an event that took place during one remarkable night in May 1943.

These are the personal testimonies of those who were involved in the raid on the dams, and these are their words – I have been but a catalyst.

Max Arthur
London, 2008

CONTENTS

Foreword

I have been a fervent admirer of Max Arthur for some years now. His voice has more authority than a hundred other historians because it is a voice that is almost always silent. How common it is for us to pay lip service to the idea that history can be better understood by listening to the men and women who lived through it than by reading the abstract judgments of historians sitting at desks, but how easy it nonetheless is to fall into the trap of becoming one of those judges, shrewdly, gravely, astutely assessing strategy and admonishing policy from the safe distance of the present. Into this trap Max Arthur never falls; he never intrudes on the lively witness of those who were there. His business is the scrupulous sourcing, selection and presentation of written and spoken testimony and its presentation to us uncontaminated by dogma, doctrine or theory.

I have recently had the good fortune to meet some of the very few living participants in 617 Squadron's legendary 1943 raid on the Ruhr dams: one of the highlights of my life was taxiing in a Lancaster bomber with Ray Grayston, flight engineer on Leslie

Knight's AJ-N 'Nut', the plane that destroyed the Eder Dam. I was writing a screenplay on the Dambusters for the film-maker Peter Jackson. Like almost any Briton, Australian or New Zealander, I had been brought up on Michael Anderson's masterly 1954 film. I am one of those who cannot hear the Eric Coates's 'Dambusters March' without tears pricking my eyes. The combination of imagination, engineering brilliance, obstinacy and determination that lay behind Barnes Wallis's development of 'Upkeep', the bouncing bomb (which we now know was a bouncing and *spinning* bomb), together with the dedication, fortitude, daring, skill and bravery of the nineteen seven-man bombing crews who deployed that bomb makes for a story that will be told for all time. But it should be told *right* and Max Arthur knows how that is done.

The bomber crews of World War Two suffered great attrition, but they suffered neglect and silence too. Those who were not there have been all too keen to tell us how wicked the Allied bombing policy was, and the dams raid itself has been written off by some as a waste of men and resources. Even if all this scepticism were justified – and I do not believe that it is – collective guilt, shame and disapproval of political and military aims and means should not alter our view of the remarkable expertise and breathtaking courage of the crews themselves.

Reading this book teaches one a valuable lesson. All the voices, in their humour, self-deprecation and unflinching honesty confirm that the heroism underlying the actions of the 617 Squadron crews was the kind of quiet heroism that consisted of doing a job, a job that was fearsomely technical and difficult and undertaken in appallingly cramped and uncomfortable conditions. Being a part of that famous raid was not about being flamboyant and boisterous or being wittily laconic into the intercom, it was about practice, practice, practice (for what they knew not). Then, on the day, it was about the constant monitoring of data – glide paths, magnetic compass deviations, dead reckoning pinpoints, calculations of fuel according to atmosphere and altitude and so on. These men were not just brave beefy chaps; they had real brains. Lancasters cannot take off at night in formation and fly low for hundreds of miles, drop an enormous bomb that is spinning at 500 revolutions per minute from exactly the right height, at exactly the right speed and exactly the right distance from the target and then move on to *another* target before returning home – all the time under fire from enemy anti-aircraft batteries – without a particular kind of steady, unblinking courage, tenacity and will that is out of the ordinary, so much out of the ordinary in our age that it might now be said to be extinct.

What a story. And I do not believe that it has ever been better told.

Stephen Fry

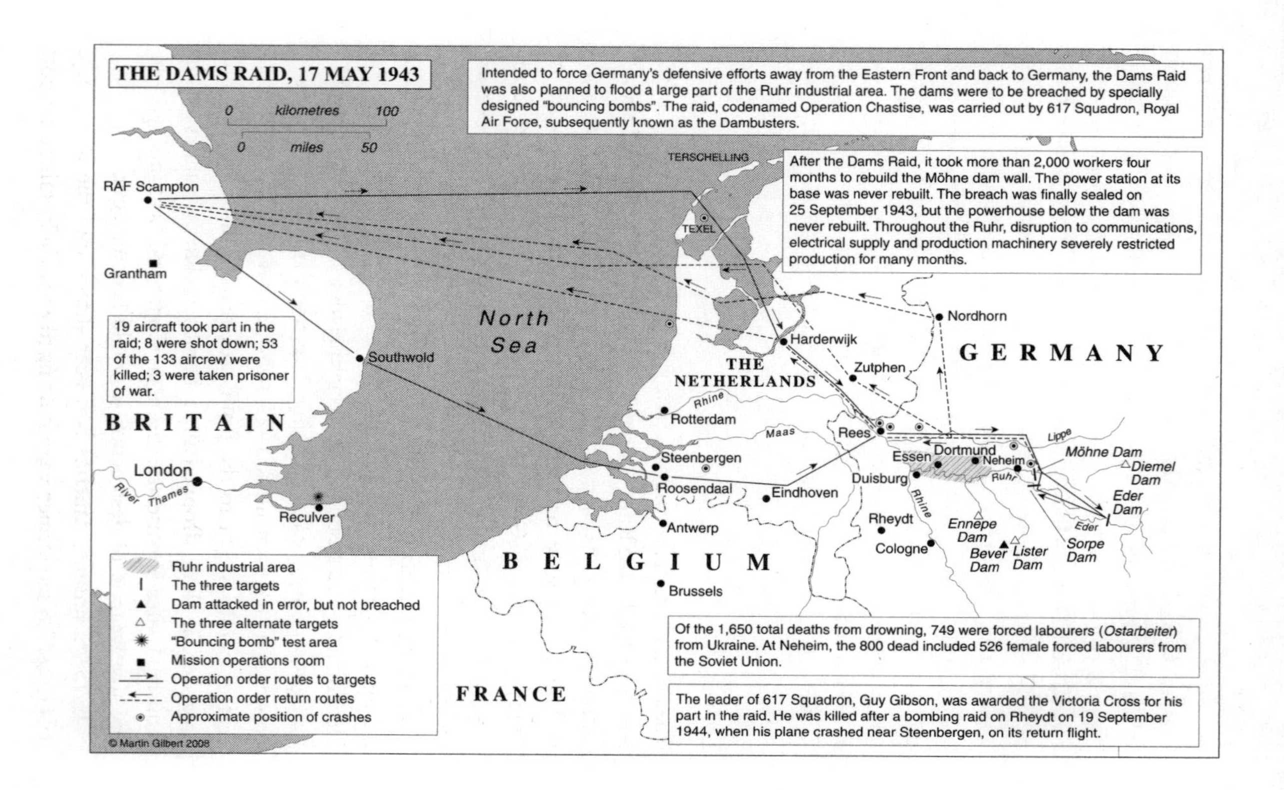
THE DAMS RAID, 17 MAY 1943
0 kilometres 100
0 miles 50
Intended to force Germany's defensive efforts away from the Eastern Front and back to Germany, the Dams Raid was also planned to flood a large part of the Ruhr industrial area. The dams were to be breached by specially designed "bouncing bombs". The raid, codenamed Operation Chastise, was carried out by 617 Squadron, Royal Air Force, subsequently known as the Dambusters.
After the Dams Raid, it took more than 2,000 workers four months to rebuild the Möhne dam wall. The power station at its base was never rebuilt. The breach was finally sealed on 25 September 1943, but the powerhouse below the dam was never rebuilt. Throughout the Ruhr, disruption to communications, electrical supply and production machinery severely restricted production for many months.
19 aircraft took part in the raid; 8 were shot down; 53 of the 133 aircrew were killed; 3 were taken prisoner of war.
RAF Scampton
Grantham
TERSCHELLING
TEXEL
North Sea
Southwold
BRITAIN
London
River Thames
Reculver
Harderwijk
Nordhorn
GERMANY
THE NETHERLANDS
Zutphen
Rhine
Rotterdam
Maas
Rees
Lippe
Steenbergen
Roosendaal
Eindhoven
Essen
Dortmund
Neheim
Ruhr
Möhne Dam
Diemel Dam
Eder Dam
Eder
Duisburg
Rhine
Rheydt
Cologne
Ennepe Dam
Bever Dam
Lister Dam
Sorpe Dam
Antwerp
BELGIUM
Brussels
FRANCE
Ruhr industrial area
The three targets
Dam attacked in error, but not breached
The three alternate targets
"Bouncing bomb" test area
Mission operations room
Operation order routes to targets
Operation order return routes
Approximate position of crashes
Of the 1,650 total deaths from drowning, 749 were forced labourers (*Ostarbeiter*) from Ukraine. At Neheim, the 800 dead included 526 female forced labourers from the Soviet Union.
The leader of 617 Squadron, Guy Gibson, was awarded the Victoria Cross for his part in the raid. He was killed after a bombing raid on Rheydt on 19 September 1944, when his plane crashed near Steenbergen, on its return flight.
© Martin Gilbert 2008

INTRODUCTION

The concept of an air raid to destroy the dams of Germany's industrial heartland – the Ruhr district – was no sudden whim by Britain's military planners. As early as 1937, as Hitler's Reich-building activities and aggressive rearming policy sent waves of alarm across Europe, the British Government started looking at ways to defeat a new and rearmed German military machine if war became inevitable.

In 1939, just days after the start of the war, the RAF launched daylight bombing attacks on Wilhelmshaven and, later, on Heligoland – and sustained unacceptable losses. The policy changed to high-level, night-time bombing, targeting major industrial areas; however, an assessment in 1941 showed that only about ten per cent of bombers sent to the Ruhr area reached their target – and only one in three dropped their bombs within five miles of the industrial sites themselves. Bomber Command was not yet the effective weapon of destruction it would become later.

In the meantime, America had been drawn into the war in late 1941, and US servicemen were starting to arrive in Britain to support the Allied offensive in Europe. A change in attack came with the appointment in February 1942 of Air Marshal Arthur 'Bomber' Harris as Commander-in-Chief of Bomber Command. He was not averse to mass raids on cities and, if necessary, the destruction of civilian targets – indeed, Lord Cherwell, one of Churchill's foremost scientific advisers, drew up a list of fifty-eight German cities which, if destroyed by attrition bombing, would bring the Reich to its knees. The first thousand-bomber raid was launched against Cologne on 30 May 1942. But one thing remained certain, high-level bombing of industrial targets was costly in both men and machines. German efficiency meant that factories were quickly rebuilt and plants dispersed over wider areas, making them even harder to target. With little alternative, however, Harris persisted with his bombing policy. Events on the Eastern Front, however, gave the Allies a whiff of hope that the tide might be turning in their favour. In June 1941, Hitler launched an invasion of Russia, but by the end of 1942, his army had become bogged down in Stalingrad where the fierce determination of the defending Soviet troops and the cruel winter conditions combined to bring the invading army to its knees. A decisive strike against the weakened Reich might mark a turning point in the European theatre too.

When the idea of targeting the Ruhr dams had first been discussed in 1937, it was abandoned because such massive structures appeared to be impossible to break, but all the same, various scientists started independent research into ways to crack the dams and bring German war production to a standstill by depriving industry of the water supplies essential for steel manufacture.

Their problem was to establish how much explosive would be required to breach the dams and then find a way of placing it accurately on the target. If they were to succeed they would require the full co-operation of the RAF and the best scientific minds in the country – and with the risks inherent in any such exploit, the military commanders would have to be convinced that the chances of success and the ensuing level of damage would justify the risk to human life.

After many meetings and presentations by eminent scientists and inventors, it was decided to go ahead with a plan to bomb the dams using a revolutionary weapon developed by the genius and tenacity of Barnes Wallis.

Chapter 1

Forming the Squadron

After the decision had been made in February 1943 to mount a raid against the Ruhr dams, there was a race against time to form and train a squadron to carry it out by the mid-May deadline. This, it was calculated, would be the time when the reservoirs were at their fullest and the pilots could rely on the light of a full moon to help them bomb on target.

The new squadron would have to be made up from men of 5 Group of Bomber Command, flying Lancasters, the only aircraft capable of carrying the necessary bombload. Flying standards and skills would have to be of the very highest, so the best choice would be experienced, ideally tour-served crews, even though many were due a well-earned break. As the chiefs of Bomber Command considered the tough new appointment, one man, the vastly experienced twenty-four-year-old Wing Commander Guy Gibson, DSO and bar, DFC and bar, stood out as the choice to lead the new squadron. He had developed the already high-performing 106 Squadron into the finest unit in 5 Group – he was always looking for new ways to improve operational effec-

tiveness and was the first to have all his aircraft equipped with cameras to assess bombing accuracy.

At the end of his tour with 106 Squadron in March 1943, and ready for some leave with his wife in Cornwall, Gibson was surprised to be posted to 5 Group HQ on 15 March. His first impression was that he was to help with the writing of a book, and he kicked his heels for three days. In the meantime Air Marshal 'Bomber' Harris instructed Air Vice-Marshal the Honourable Ralph Cochrane, Air Officer Commanding 5 Group, to form a special squadron under Gibson for a raid against the Ruhr dams. On 18 March, Cochrane met with Gibson to sound out his willingness for one more operation, and the day after, Gibson was tasked with bringing together a crack bomber squadron and training them for a top-secret mission – for which not even he knew the target. Despite Gibson's claim that he had 'picked them all myself', this was not correct. A cadre of the captains and aircrew were known personally to Gibson, but others appear to have been recommended to him, or selected for him – this is supported by the fact that some were soon posted out from 617 once training began.

Lancasters for 'Operation Chastise' were delivered to Scampton – coded 'AJ' plus a serial letter. Gibson's aircraft was 'G' for George, Shannon's 'L' for Leather, and so on.

Wing Commander Guy Gibson

I had been at Group Headquarters, Grantham, one or two days and had tried to get down to the factual business of writing, when the AOC sent for me. Air Vice-Marshal Coryton had gone, to the deep regret of everyone in the Group, and the new Air Vice-Marshal was the Honourable Ralph Cochrane, a man with a lot of brain and organising ability. In one breath he congratulated me on my bar to the DSO, in the next he suddenly said, 'How would you like the idea of doing one more trip?'

I gulped. More flak, more fighters; but said aloud, 'What kind of a trip, sir?'

'A pretty important one, perhaps one of the most devastating of all time. I can't tell you any more now. Do you want to do it?'

I said I thought I did, trying to remember where I had left my flying kit. He seemed to be in such a hurry that I got the idea it was a case of take-off tonight.

But two days went by and nothing happened. On the third he sent for me again. In his office was another man, one of the youngest Base commanders in the Group. Air-Commodore Charles Whitworth. The Air Vice-Marshal was very amiable. He told me to sit down, offered me a Chesterfield and began to talk.

'I asked you the other day if you would care to do another raid. You said you would, but I have to warn you that this is no ordinary sortie. In fact, it can't be done for at least two months.'

I thought 'Hell, it's the *Tirpitz*. What on earth did I say "Yes" for?'

'Moreover,' he went on, 'the training for the raid is of such importance that the Commander-in-Chief has decided that a special squadron is to be formed for the job. I want you to form that squadron. As you know, I believe in efficiency, so I want you to do it well. I think you had better use Whitworth's main base at Scampton. As far as aircrews are concerned I want the best – you choose them. Wing Commander Smith, the SOA, will help you pick ground crews. Each squadron will be forced to cough up men to build your unit up to strength.'

Cochran went on, 'Now, there's a lot of urgency in this, because you haven't got long to train. Training will be the important thing, so get going right away. Remember you are working to a strict timetable, and I want to see your aircraft flying in four days' time. Now you go upstairs to hand in the names of your crews to Cartwright; he will give you all the help you want.'

'But what sort of training, sir? And the target? I can't do a thing –'

IWM CH13618

Wing Commander Guy Gibson, at twenty-four already a vastly experienced bomber pilot, who set up and trained the air crews of 617 Squadron.

'I am afraid I can't tell you any more just for the moment. All you have to do is to pick your crews, get them ready to fly, then I will come and see you and tell you more.'

'How about aircraft and equipment?'

'Squadron Leader May, the Group Equipment Officer, will do all that. All right, Gibson.'

He bent down to his work abruptly. This was a signal for me to go.

Flight Lieutenant David Shannon

PILOT, AJ-L

I was trained and commissioned as a pilot into the Royal Australian Air Force. After arrival in the UK I trained on Whitleys and was posted to 106 Squadron, commanded by Guy Gibson. I did some thirty-seven or thirty-eight ops with 106, then they wanted to take me off flying and send me for training. I didn't like that, so I applied to go on to Pathfinders and went off to 83 Squadron. I'd only been there twenty-four hours when I got a call from Guy Gibson.

'I'm starting up another squadron for a special raid,' he said. 'I can't tell you what it is, but if you'd like to join me again, I'll be only too willing to have you back.'

I said, 'Yes.'

I arrived to join Gibson's new squadron at the end of March 1943, and as it was a brand-new squadron, we had no equipment, no aeroplanes – nothing. So the first thing was to get the whole thing going, which was a mammoth job for Gibson to do from scratch. We hadn't even been designated as a squadron with a number at that time, and it was just known as Squadron X for quite a long time. Gibson was given carte blanche, and he personally, with the aid of a Senior Air Staff Officer from 5 Group, chose all his crews that he wanted to form this new special squadron.

I'd been selected from 106, and a great friend of mine, John Hopgood, known as 'Hoppy', came too, along with another pilot – a Canadian, Flight Sergeant Burpee. I think we were probably the first three that Gibson chose, because he knew us – then he went through the records of other aircrew and selected the squadron until he'd built up a force of two flights – twenty-one crews in total – including himself, as the squadron commander and then two flights of ten.

By the time we got there, all the selections had been made by Gibson and the people assisting him in 5 Group Headquarters. By the end of the first week in April the squadron had been formed – twenty-one crews of seven men each – so there were 147 men as aircrew and supporting ground staff – mechanics, fitters, administrative staff. So there were some 500 as ground staff. He had a mammoth operation to undertake in such a short space of time, but he was given full assistance. Orders had come down from Bomber Command through 5 Group to get on with this, because time was short. Nobody knew why time was short, but that was the order.

Wing Commander Guy Gibson

It took me an hour to pick my pilots. I wrote all the names down on a piece of paper and handed them over to Cartwright. I had picked them all myself because from my own personal knowledge, I believed them to be the best bomber pilots available. I knew that each one of them had already done his full tour of duty, and should really now be having a well-earned rest; and I also knew that there was nothing any of them would want less than this rest when they heard that there was an exciting operation on hand.

We would also require ten aircraft to begin with, with all their gear. Later we would need more. This was a big job; there were trestles, trolleys, spare wheels and bumble motors. Cartwright knew his job inside out – he promised that they would be delivered at Scampton the next day.

Next morning to the personnel officer, to fix up the ground staff. We were taking a few ground crews from each squadron, but in the case of the NCOs it would be necessary to have the very best men available. I asked for them and got them. Then along to the WAAF officer to see that we got our fair share of MT drivers and cooks – very important.

By now things were beginning to get beyond me. I went to the stationery department and got a little book and wrote down everything to be done in a long column. Every time anything got fixed up, I would tick it off, but by the end of the day, there weren't many ticks to be seen. Then would come the visit to the Senior Air Staff Officer, the Air Officer Commanding's deputy. This was Air-Commodore Harry Satterly, a big, blunt man who had the habit of getting things done quickly and well. His help was invaluable – in fact, I don't know what could have happened without him.

So, by the end of two days the squadron was formed. It had no name and no number. We had worked too fast for that branch of the Air Ministry which gives squadrons new numbers and identification letters, and we decided to call it simply Squadron X.

Flight Sergeant Ken Brown

Pilot, AJ-F

I was flying with 44 Squadron. My C/O was a VC winner, Wing Commander John Nettleton, who had led the Augsburg raid in April 1942, and we were briefed to go to Berlin. After the briefing he said, 'Brown, report to my office immediately after the briefing.' Which I did and he said, 'You are transferred to a new squadron.'

I wasn't too happy about that. I said, 'Sir, I'd rather stay here and finish my tour with Forty-four.' He explained in his very curt manner that this was impossible. It was a name transfer and he could do nothing about it.

So we went to Berlin and on our return we got packed up and off we went to No. 617. But before we went, the Wing Commander wished me well and said, 'Do you realise Brown, you're going to be the backbone of this new squadron.'

We arrived over at Scampton and started to look around as to who was there. There were an awful lot of DFCs, not so many DFMs. We realised that perhaps we weren't really all what we were set up to be.

My wireless operator sauntered up to me and said, 'Skip, if we're the backbone of this squadron, we must be damn close to the ass end.' I began to wonder how I'd got there.

When I was going through Manchester training and Lancaster training there was a fellow by the name of 'Mick' Martin who perhaps was to become one of the RAF's greatest. He was my instructor at that time. So was a fellow that we knew as 'Terry' Taerum (who became Guy Gibson's Navigator on 617). Everyone in the outfit knew Terry. He was teaching GEE at the time. GEE was a navigation aid. It was new at the time, so Terry was sort of our expert.

Sergeant Ray Grayston

Flight engineer, AJ-N

I stayed with Les Knight and we completed a tour of thirty ops – I think I was about two short, but they agreed I could count that as a complete tour to stay with the crew. At the end of that we were due to be stood down, but they approached us and asked would we stay together to do one more trip. We agreed to that – we had a very good crew. Then we were told to go with our own aircraft and our ground crew and go to Scampton. We arrived and there was not a lot of information available – everybody was new – we knew our own crew but some were teamed up as new crews. We agreed as a crew we'd stay and do whatever they wanted us to do.

Flying Officer Harold Hobday

NAVIGATOR, AJ-N

The first squadron I was on was 50, and we'd almost finished our tour – and not many people finished tours because, I'm afraid, the loss rate was so colossal. I decided I'd like to go on a special navigation course, and I was duly put up for this – then we were invited to go on a special squadron as a crew. And much to the dismay of the person who put me up for this special navigation course, I decided that I'd rather go with my crew to this new squadron. We didn't know what it was all about, but we thought it was something special. We had done quite well on 50 Squadron, and I think that's why we were chosen as a crew to go to 617 Squadron.

I opted for the new squadron because I didn't want to let my crew down, and I was quite keen on bombing. I loved the life. It may sound terrible now in peacetime to think you liked bombing people, but I liked the idea of the crew staying as one integral part of the set-up. I wouldn't have liked the thought of another navigator taking my place in my own crew. We were strictly volunteers, but the CO of our old squadron put it to us. Gibson had something to do with it – and our record must have been the main reason why we were chosen.

I do remember that I spent a sleepless night worrying whether I should go on this special squadron or go on the course – because one doesn't like to ask somebody to put you on a course, and when they've arranged everything, to say, 'No, I'm sorry but I'm not going.' But I had to make this choice, and I spent a lot of time worrying about what to do for the best. But I decided the crew should come first, and I would stay with them rather than go on the course – which would have meant promotion for me.

We were all keen to find out what it was – but we didn't find out until just before the raid exactly what we were going to attack – but we understood it was something really special and it was rather nice to be involved in something that special.

Flight Lieutenant Les Munro

Pilot, AJ-W

I'd trained as a pilot and reached 97 Squadron in December 1942, and served there until March '43. At that stage myself and crew had done around twenty-one trips, and 5 Group HQ called for volunteers from crews nearing the end of their first tour or commencing their second. I had a discussion with my crew and we decided that we would volunteer – although we didn't know what we were volunteering for, other than that it was for a special operation and a squadron was being formed for that purpose. So, if they wanted volunteers, we were prepared to volunteer.

The crew were reasonably happy about it. Nobody expressed any particular concern or opposition to volunteering for a special duties squadron. I wouldn't have taken the crew with me if they hadn't been happy. But it was a natural assumption that it was going to be a difficult job. Looking back I think it was just a question of, 'OK, they're looking for volunteers for something. Here we are nearly at the end of our first tour. Let's have a go.' But it was a matter of conjecture as to what we were volunteering for.

My first bomb-aimer had started passing out at high levels, so I'd had a succession of bomb-aimers. And when I arrived at Scampton, I'd got another bomb-aimer, Jimmy Clay.

Flight Sergeant Robert Kellow

WIRELESS OPERATOR, AJ-N

The offer presented to us sounded interesting, and with our faith in each member's ability, we made up our minds there and then that we would accept the offer and move across as a crew to this new squadron.

Sergeant George 'Johnny' Johnson

BOMB-AIMER, AJ-T

617 Squadron was formed in March '43 and the American, Joe McCarthy, the captain, was asked if he'd join it – it was to be a squadron of all experienced crews. It was scheduled as a special squadron for a special trip – and it was stressed at that stage *a* special trip. It consisted of twenty-one experienced crews, all of whom had done one tour or fast approaching it – some had done two.

Flight Sergeant George Chalmers

WIRELESS OPERATOR, AJ-O

At the time when 617 Squadron was brought together, I was stationed at RAF Abingdon and we were doing blind flying – flying approaches – and I got a little bit bored of this job and I think my squadron commander got a bit tired of me – so between us we agreed I'd go back on operations and 617 just happened to be the squadron. So it was quite by chance on my part, other than the fact that I wanted to go back on operations.

When I arrived at Scampton, I was crewed up with Flight Sergeant Townsend, in an NCO (non-commissioned officer) crew – which I asked for at the time. It wasn't so much a selection as an agreement. I was an NCO myself and I always had an affinity with sergeant pilots. I felt you had more comradeship with a chap of your own rank. You'd live with them and you messed with them as well. That was the main reason, the comradeship, I think.

When I got to Scampton the Flight Sergeant took my particulars. 'Report to the Intelligence Officer straight away. Leave all your stuff here.'

'Oh, that's unusual. I've never done that before. Can I book in?'

'No, go and see him.'

So I went and saw this fellow in the headquarters. He sat me down and said, 'Read that lot', and it was the Official Secrets Act. I had read them before several times over, so I was doing this and I suppose about fifteen minutes went by.

'Have you read them?' he said.

'Well, I have read most if it already many times,' I said.

'That is not the point,' he said. 'Have you read all of it?'

'No, not really,' I said.

'Well, sit down,' he said, so I did, and read this bloody thing from back to end. I took half an hour and for God's sake I'd had enough.

'Sign here,' he said. So I signed and am going down the stairs, and who should be walking up but the Station Commander. I could see all this gold braid, and when he lifted his head, it was bloody Whitworth. I knew him, and we chatted for a bit on the stairs, and I told him I was joining.

Life after that was just one hectic thing after another.

I was supernumerary, being posted on my own. Most of the crews were coming in as complete crews – although I didn't know this – but Bill's wireless operator didn't come with him because he was getting married, so he brought the crew minus a wireless operator. I was sitting in the crew room and talking to a navigator from down London way. Just then Micky Martin came in.

'I haven't got you crewed up yet,' he says. 'You have got a choice – and who would you like to fly with?'

'Well anybody,' I said.

'Have you any preferences?'

'I'd prefer to fly with an all NCO crew if I could,' I said.

He was a bit taken aback and said, 'Well, I think we can fit you up.' The next thing I know, Bill Townsend came aboard and joined me in conversation, and that was it.

Sergeant Dudley Heal

NAVIGATOR, AJ-F

In Bomber Command you do a tour of operations of thirty trips – that is your first tour. When you have done your thirty, you are rested for six months and then you come back and do a second tour of twenty trips, after which you can, if you so wish, say, 'Right, I'm not doing any more operations.' So when we landed, at the end of our seventh trip I think it was, and at our debriefing were told the next day we were being transferred to another station – another squadron – we were absolutely flabbergasted. No reason was given – we were just told that we were being posted the next day to Scampton.

There's a new squadron being formed. It hadn't got a number – we didn't know anything about it. But that was it – we had to go. By 2 April, we were at Scampton. There were a lot of officer air crews on this squadron – there were some twenty crews who had recently arrived and a good number of these officer pilots were already highly decorated – which aroused certain interest and also made us wonder what sort of operation we were going to be in for.

Flying Officer Edward Johnson

BOMB-AIMER, AJ-N

I volunteered for the RAF and was accepted to train as an observer, which was a general category then for navigation, gunnery, wireless – but later split into navigators, bomb-aimers and separate trades. So then I trained as an observer and got my flying 'O' badge.

I met up with this crew at 50 Squadron, some of whom I knew, and they needed a bomb-aimer, so with Group's permission I went flying as a bomb-aimer.

We'd just finished a tour of ops with them and we had a good reputation for getting there and getting back – and we were more or less invited to join 617 Squadron. Everybody in the crew was in agreement to going because we didn't want to split up – which is what happened if you finished a tour and had to go on training duties, you'd probably come back with a new crew – which none of us fancied at all. It seemed a good move to keep on flying together.

Initially nobody knew what the role of 617 Squadron was to be. There were all sorts of rumours flying about – about it being for special duties – but nobody had the vaguest idea what it was actually going to do. All the people joining it were very experienced pilots – mostly decorated or having done one or two tours of operations, so it was obviously something rather special to have all these well-known bods.

Flight Sergeant Leonard Sumpter

Bomb-aimer, AJ-L

I'd done my navigational training in Canada and my operational training at Bassingbourn, then I was assigned to 57 Squadron, based at Scampton, where I did thirteen operations. Then my pilot, Flight Lieutenant George Curry, was grounded with ear trouble and would be down for a long time – which meant the crew breaking up. In the next two hangars a new squadron was forming – 617 Squadron. But it wasn't 617 – nobody knew the number of it then. I and Lofty Henderson, the flight engineer, heard that the fellow named Shannon was looking for a bomb-aimer and a flight engineer, so we arranged an informal meeting with him in the hangar. We looked him over and he looked us over – and that's the way I got on to 617 Squadron. There wasn't all this selection of crews – that's a load of eyewash. Gibson might have picked a few pilots that he knew – but the crews themselves just got together. We didn't get a wireless operator until almost the end of April.

One or two pilots brought their own crews with them – but the rest of it, it was just getting together and finding a pilot – or him finding you. You crewed up and as long as you got on together and were amenable to each other, that was it. Certainly no interviews – and I only ever spoke to Wing Commander Gibson once in my life – which was when he tore me off a strip.

Flight Sergeant Ken Brown

PILOT AJ-F

The squadron was being formed at Scampton, but they didn't have any aircraft – they didn't even have any parachutes. They were gracious enough to say that we could stay in some of the old married quarters until we found out where we belonged.

Wing Commander Guy Gibson

On the evening of 26 March I arrived in the mess at Scampton with Nigger, my dog, sniffing his way along happily at my heels. I think he smelled a party. He was right. In the hall were the boys. I knew them all – that is, all the pilots, navigators and bomb-aimers – but there were a few strange gunners and wireless operators knocking around.

Within a second a whisky was shoved into my hand and a beer put on the floor for Nigger, who seemed thirsty. Then there was a babbling of conversation and the hum of shop being happily exchanged; of old faces, old names, targets, bases and of bombs.

Courtesy of www.ww2images.com.

Guy Gibson tempting Nigger with another pint.

This was the conversation that only fliers can talk, and by that I don't mean movie flyers. These were real living chaps who had all done their stuff. By their eyes you could see that. But they were ready for more. These were the aces of Bomber Command. All in all, they probably knew more about the art of bombing than any other squadron in the world.

Playing with Nigger and trying to get him tight were some of my own crew – Terry, Spam, Trevor and Hutch had long since gone to bed with the rest. Nigger knocked back about four full cans, made a long, lazy zig-zag trail of water down the corridor, then went to bed himself with his tail between his legs.

Flight Sergeant George Chalmers

BOMB-AIMER, AJ-O

It was more like a Commonwealth squadron with Canadians, Americans, New Zealanders – it was different from any squadron I'd ever been on before.

John Elliott

NCO GROUND CREW

Scampton was good – it was a peacetime station, with permanent buildings. It was a bit overcrowded when we got there. Very few of us senior NCOs could get into the Sergeants' Mess to be billeted, so we got billeted in what, in peacetime, were the married quarters. In our time off we went to just about every pub in Lincoln – but strangely enough we were complimented afterwards on the way we had maintained our security. As far as I am aware there were only two breaches of security but nothing really bad. Before we started training we all had a big meeting and were told it had to be secure – there would be no talking about anything – nothing written in letters written. As it was, all my letters home were censored.

Flying Officer Edward Johnson

Bomb-aimer, AJ-N

Scampton was a typical pre-war aerodrome with good hangars and good accommodation – and a nice mess. The only drawback to it was that it was a grass field with no runways. But certainly it was a first-class aerodrome.

We were a mixed crew of young and older men – of practically all nationalities. The pilot was the Australian, Pilot Officer Les Knight, and very young – just twentyish. The rear gunner and mid-upper gunner were both Canadians and very young too – one nineteen and the other twenty or so. The wireless operator was also Australian and the navigator and I were British – and we were much older than the others in the crew. But there was never any difficulty about that. Relations among the crew were excellent and all of us very mixed characters got on very well together. Les Knight didn't drink or smoke and was a bit religious, while the two Canadians liked to drink, and so did I – and the others. But the pilot always went with us and had a lemonade, and this proved important as time went on. The pilot, navigator and bomb-aimer were all commissioned and the rest were sergeants, but we all mixed socially.

Flying Officer Harold Hobday

Navigator, AJ-N

I reckon we were a first-class crew. The CO of our previous squadron, 50, used to call us 'the Ace Crew'. The mid-upper gunner was an absolute ace at shooting – but we were all very dedicated – and all very young except the bomb-aimer, Johnson, and myself, who were the oldest in the crew. I was thirty-one, as was he, so we were like the granddads of the squadron, because thirty-one was pretty old for aircrew in those days.

We were quickly taken over to Scampton and met Gibson – had a bit of a mess party that night – the first night we were there. After that we knuckled down to some very serious training.

Sergeant Ray Grayston

FLIGHT ENGINEER, AJ-N

Les Knight was an amazing fellow – he'd been a student and trained to fly in South Africa, I think, then came to England. He was a very quiet chap, he didn't drink and he didn't go out with women, couldn't ride a bike or drive a car, but he could fly a Lancaster. Amazing chap and dedicated to his work, he always completed the job we were given, no matter how hot it got, with people screaming, 'Drop the bomb' – he'd never do anything until we got to the target.

Sergeant Frederick Sutherland

FRONT GUNNER, AJ-N

Our pilot, Les Knight, was short, very muscular, strong in the shoulders and arms. He was a good disciplinarian. He was very quiet, but if you were out of line, he quietly told us that you know you'd better not do that again. So we didn't. We respected and admired him. He was just a wonderful person.

John Elliott

NCO GROUND CREW

When we started getting some of the crews in, we began to think there was something more in the wind. We couldn't see it being a normal squadron because of the level of experience of the people who were coming to us. They were all very experienced as an aircrew. Some of them had already done two operational tours, so we got the idea that we were going to be something a bit special.

Sergeant Jim Clay

BOMB-AIMER, AJ-W

We did manage to put our flight offices into some sort of shape by putting up blackboards with rawlplugs and scraping old tables clean with bits of broken glass.

Wing Commander Guy Gibson

Jim Heveron, 'Boss of the orderly room', had to beg, borrow and sometimes steal from every imaginable source the much required bumph, so that he could start some sort of filing system.

Harold Roddis

FLIGHT MECHANIC

No sooner had I set foot in the main gate than the Service Police collared me with torrents of verbal abuse. They said I didn't measure up to what was expected of a smart airman and put me on a charge. But I was used to working in dirty, greasy overalls which stank of burned engine oil and 100 octane fuel. In winter we wore our working blues under the overalls. People had been known to get up from the mess table and move away from us. However, Gibson quickly stepped in and had all the charges lifted.

Flight Lieutenant Harold (Harry) Humphries

ADJUTANT, 617 SQUADRON

Trouser removal was the standard thing to be initiated into the squadron. One time I heard the words in the mess, 'How about we take the Adjutant's trousers off?' I didn't fancy this, so I climbed out the window and shinned down the drainpipe. When they heard about this I became something of a hero for a couple of days – so I didn't lose my trousers.

Sergeant George 'Johnny' Johnson

Bomb-aimer, AJ-T

We had a gunner on the squadron, Buckley, who said that he reckoned he could drink Joe McCarthy under the table. He was five feet plus, Joe was well over six and built in the same sort of way, outwards as well as upwards. So, one non-operational session they had a competition. It started in the bar at lunchtime and after the bar had closed they had their little kip in the afternoon and came down for bar opening at six o'clock in the evening, and at half past seven Joe carried him to bed.

Wing Commander Guy Gibson

Joe McCarthy was one of the Americans who had volunteered to fly in the RAF before America was at war with the Axis. He had been given the chance to rejoin the United States forces, but had preferred to stay with the boys.

Flight Lieutenant Les Munro

Pilot, AJ-W

There were no outstanding personalities in the crew – the wireless operator and the rear gunner were Canadians – rather hard cases in some respects. My navigator, who I would say was outstanding in his field, was a Scotsman – a bloke called Jock Rumbles. The other three were Englishmen. We knitted pretty well as a team – I always regarded myself as fortunate to have a team of the calibre that I had.

Wing Commander Guy Gibson

I got them all together. There were twenty-one crews, comprising 147 men; pilots, navigators, wireless operators, bomb-aimers, flight engineers. Nearly all of them twenty-three years old or under, and nearly all of them veterans. I saw them in the same old crew room, which brought back many memories of 1939–40 days. Now it was packed full of young, carefree-looking boys, mostly blue-eyed, keen and eager to hear the gen. I felt quite old among them.

My speech to them was short. I said, 'You're here to do a special job. You're here as a crack squadron – you're here to carry out a raid on Germany which, I am told, will have startling results. Some say it may even cut short the duration of the war. What the target is, I can't tell you. Nor can I tell you where it is. All I can tell you is that you will have to practise low flying all day and all night until you know how to do it with your eyes shut. If I tell you to fly to a tree in the middle of England, then I will want you to bomb that tree. If I tell you to fly through a hangar, then you will have to go through that hangar, even though your wing-tips might hit either side. Discipline is absolutely essential.

'I needn't tell you that we are going to be talked about. It is very unusual to have such a crack crowd of boys in one squadron. There are going to be a lot of rumours – I have heard a few already. We've got to stop these rumours. We've got to say nothing. When you go into pubs at night, you've got to keep your mouths shut. When the other boys ask you what you're doing, just tell them to mind their own business, because of all things in this game, security is the greatest factor.'

In one of the big hangars, Chiefy Powell got all the ground crews together and I stood on top of my Humber brake and spoke to them. I said much the same as I said to the aircrews, stressing that security was the most important thing of all. There was to be no talking. Then Charles Whitworth got up and welcomed 617

Squadron officially to Scampton Base. He made a good speech, and I remember thinking that it would be a good idea if I could remember it so as to copy him on some future occasion, but now I have forgotten what he said except for one thing.

He said, 'Many of you will have seen Noël Coward's film, *In Which We Serve*. In one scene, Coward, as commander of a destroyer, asks one of the seamen what is the secret of an efficient ship. The seaman answers, "A happy ship, sir." And that is what I want you to be here. You in the Air Force use a well-known verb in practically every sentence; that verb is to bind. I can promise you that if you don't bind to me, I won't bind to you!'

Sergeant Jim Heveron

IN CHARGE OF THE ORDERLY ROOM

The aircrew and ground crew personnel were all aware that they were training for a special op and that made the general feeling one of urgency. I believe it is the shortest time in the annals of the RAF that a large bomber squadron assembled, received twenty-one aircraft, serviced them and trained the crews to fly such a specialised operation.

CHAPTER 2

Wing Commander Guy Gibson

In choosing Guy Gibson to command the new special-mission squadron, the RAF top brass had found a man of exceptional professionalism and enormous experience. Already decorated four times, twenty-four-year-old Gibson was a model of efficiency and patriotism – but he was not universally popular. He could be a stickler, a bit of a stuffed shirt and mingled socially only with his fellow officers – although no-one could question that he was the man for the job.

John Elliott

NCO GROUND CREW FOR GIBSON

I thought the squadron leadership was quite good. Our CO at 106 was Guy Gibson, who eventually formed 617. He was with us at 106 when he was asked to form this new squadron. He seemed quite reasonable when he first arrived, and I got to know him better when 617 was formed. I thought he was very fair and reasonable – and he understood our problems as we tried to understand his.

Courtesy of The National Archive, ref. INF2-43 (1124).

Guy Gibson. 'I believe he was the right man for the job. He had a lot of operational background. He knew crews and knew how they felt.' Flight Sergeant George Chalmers.

As far as aircraft serviceability and that sort of thing, he was quite understanding.

Gibson was special in so far as he'd done quite a few operational trips and been successful at them. He wasn't the sort who flew by the seat of his pants. I mean he knew what he was going to do all the time.

He used to fly in his shirtsleeves – especially in summer – but I certainly wouldn't call him slovenly. I don't think he was disliked but I got the job of looking after his aircraft when 617 was formed. Nobody else would volunteer to take it on. Whether that was because he was going to be the governor I don't know. I had worked under him before and he didn't give me any hassle at all. I knew he was a fair person to work for, so I said I would take it on. It was my responsibility that his aircraft was serviceable every time he wanted to fly.

I think he used to get a bit anxious if his aircraft was unserviceable for any reason, but if you explained the reason to him he was quite understanding and all he wanted then was to get the thing done as quickly as you could. He never complained – well not to me anyway.

Flight Sergeant George Chalmers

WIRELESS OPERATOR, AJ-O

I met Gibson the first day I arrived. He was our squadron commander so I saw him all the time after that. I wouldn't say he was a difficult man, but he was a strong disciplinarian. I think probably because he was a man of small stature, he tended to be on the aggressive side rather than be complacent about anything. I don't believe he was ever complacent about anything. Everything had to be right, and he let everybody know that too. He had a lot of responsibility. I believe he was the right man for the job. He had a lot of operational background. He knew crews and knew how they felt. He was a good leader.

Flight Lieutenant Harry Humphries

ADJUTANT, 617 SQUADRON

Gibson was a tough guy and didn't suffer fools gladly. There was an instance when we had a runner – a dogsbody – and Gibson pressed his bell to fetch this runner in, and as he went out he shouted out after him 'George! Come back!' but this man left the office and Gibson went storming out after him.

'Didn't you hear me calling you?'

'No sir.'

'I said, "George, come back".'

And he said, 'My name's not George.'

And Gibson said, 'If I call you bloody George, you are George.'

That was the kind of man he was.

The first thing he said to me after I'd been installed as adjutant of 617 Squadron was this squadron would either make history or be wiped out.

Flight Sergeant Ken Brown

PILOT, AJ-F

We got to the squadron. I'd never met Wing Commander Gibson before. So this was a new experience.

We were all sitting out on the lawn in front of the briefing room when someone said, 'Briefing's ready, come on in.'

So we marched into the briefing room, which was right down on the flight line. I wasn't last in, but I did close the door. When I did so, Gibson said, 'Brown, report to my office after briefing.' Sounded familiar.

However, I couldn't believe it when I reported to his office. The adjutant met me, marched me in, and Gibson had me on charge for being late for a briefing. I thought he was kidding, but he wasn't. So he then read out the charge for being late for an operational briefing, and he asked me whether I'd take a court martial or his punishment. I said, 'I'll take your punishment, sir.'

So he said, 'Fine. You'll wash all the windows on the outside of the briefing room and the inside of the briefing room. All after duty hours.'

As we were flying about eighteen hours a day, that was really something. I wasn't going to let this really stump me. So I did it. And I did it night after night. It was one of those things.

Gibson had a very high standard for everyone and you had to meet it, and meet it on his terms. He was really a strong and staunch disciplinarian. He had been brought up in a boys' school as a head prefect. And I still think he handled things in that way. At least I thought that way after the ninety-ninth window.

Kenneth Lucas

GROUND CREW

I met Guy Gibson. He came into the hangar on one occasion because a special aerial had to be fitted to his plane, as he would be guiding them in, and it was a question of the positioning of the aerial on the plane. He came into the hangar smoking his pipe – and we told him he couldn't smoke in the hangar – that was taboo. He just accepted it and our impression of him was that he was a great guy – and a very brave man.

Harry Humphries

ADJUTANT, 617 SQUADRON

Gibson was a bit of a hard nut – some people called it arrogance. He *was* hard and had a very short fuse – and he wasn't a particularly feeling sort of personality, and I never saw him show much in the way of sentiment.

Flight Lieutenant Les Munro

Pilot, AJ-W

Guy Gibson didn't spend a lot of time during the day in training, because he was involved in discussions about the development of Upkeep, and the method of attack – so we didn't see a great deal of him, but we were conscious of his character and his attitude to people – to anybody that made a mistake in training. He was always rather caustic about that. We spent a bit of time with him socially in the mess and he used to join in there at least.

Sergeant George 'Johnny' Johnson

Bomb-aimer, AJ-T

Guy Gibson was a little man – with quite a big opinion of himself – but after all, he had done an awful lot more than any of the rest of us. I'd say he probably had the right to be a bit self-important.

Flight Sergeant George Chalmers

Wireless operator, AJ-O

He had a personality which, I would say, was zero. He stuck to himself most of the time – him and his dog, Nigger.

Leading Aircraftwoman Eileen Strawson (née Albone)

Driver for Gibson

I was the only one free when he needed a driver – and that was how it all started. I liked him very much, but he wasn't very well liked, I'm afraid. I think he was too clever for most people – but I rather mothered him, as I was a lot older than him, and we got on famously.

I enjoyed driving Gibson – he always treated me as an equal, although I believe he didn't with other people. I was only a leading aircraftwoman. One day when it was very warm, he threw his hat into the back of the car.

'I wish I could do that, sir,' I said.

'Go ahead,' he said.

So I threw my hat into the back too. When we got to Scampton, I had to book in at the guardroom. I forgot my hat and was put on a charge because I wasn't wearing it. I told Guy Gibson, who was very annoyed. He jumped out very quickly and told the corporal to scrub that charge – I wasn't very popular with the corporal after that.

Flying Officer Harold Hobday

NAVIGATOR, AJ-N

We were in the briefing room when we met him, and he told us as much about the squadron as he could. What impressed me was that afterwards when he had our mess party, he was the life and soul of the party. There was a bit of a rivalry between navigators and pilots. He came up to me and he said, 'You're a navigator, are you? I'll swap jackets with you' – so we swapped jackets in the mess. It was rather a nice touch, because it made everyone feel how friendly he was. But he was quite strict too – he wouldn't stand any nonsense. If anybody drank before they were going on a practice flight, he'd be down on them like a ton of bricks – and quite rightly too. One chap had a pint of beer before he was going on a training flight, and he was severely reprimanded. Alcohol would really affect your performance – I never drank while I was flying.

Gibson would give any offenders a telling-off – he'd do it in front of the squadron, and that made you feel about two inches high.

F/L David Shannon

F/L Les Munro

Courtesy of www.ww2images.com.

F/S George Chalmers

Flying Officer Edward Johnson

BOMB-AIMER AJ-N

I had met Gibson previously at 106 Squadron, where he was the CO. He was very efficient, very straightforward – and I think everybody liked him. He wasn't a bully or a show-off, but he liked things to be done right, and he also liked you to keep fit (sometimes a sore point) with runs round the aerodrome and so on. After a night out on the tiles these weren't always frightfully popular – but I think he was on the right lines.

He was strict about work, but he was a great mixer – not so much during the working day, but always at night – he could sup his ale with anybody.

Sergeant George 'Johnny' Johnson

BOMB-AIMER, AJ-T

I knew Guy Gibson as much as an NCO on the squadron could know him, yes. I always have to be careful what I say about Guy Gibson because he did a wonderful job on this particular occasion. He was renowned before he came to 617 Squadron as being an almost brutal squadron commander. When he came to 617, from his previous associates – they were also on the squadron – he seems to have cooled down quite considerably. One wonders if that was because he knew what he'd get out of these crews, and so he would have to treat them a little more gently.

I never got rapped over the knuckles by him – but nor did I get any praise either.

Flight Lieutenant David Shannon

PILOT, AJ-L

I'd first met Guy Gibson when I was in 106 Squadron in June 1942. He'd initiated me into operations over enemy territory. When I was flying with him I seemed to be able to read his mind and anticipate every move he expected of me as his co-pilot and flight engineer. He seemed able to relax when I was with him in the cockpit. I was just twenty and he was four years older than me. We did five operational trips together with 106 Squadron. One thing we very much shared was a hatred of the Hun.

Gibson was an absolutely fantastic character. In my estimation he was one of the finest leaders of men that I've ever met. He was the type of chap that would never ask anybody to do anything he couldn't do or hadn't done himself. He was what was known in RAF terms as an operational squadron commander – that is to say that he flew just as much operationally as his crews did – which was not always the case with squadron commanders because they had plenty of administrative work and other things to do. Many of them were there for quite a long period of time to get in their full tour of operations, but Gibson was a fantastic leader. He was very strict on duty, and one of the boys off duty, and he managed to carry that off to perfection. Everybody admired him, and having been with him over a year at that stage, I was very happy to continue serving with him. We struck up quite a close personal friendship – as one did during the war – as close as one could. He was also a really first-class pilot.

Guy Gibson was a magnificent leader, but you either loved Gibson or you were scared of him. He could be very tough. Gibson's humour was robust – very earthy. He had an eye for the ladies and was, off duty, a great boozer.

Pilot Officer Lance Howard

NAVIGATOR, AJ-O

Though we did not see a lot of Gibson, he seemed to set a standard of perfection in all our training.

Flight Sergeant Leonard Sumpter

BOMB-AIMER, AJ-L

Gibson seemed like a nice little chap – he was short, broad-shouldered, smart. I liked the way he walked about, because he moved about smartly. He didn't slouch and he always looked well-dressed – trousers pressed, a clean collar on and well-shaven – and he liked a pint now and again.

As far as I was concerned, he was an ideal bloke for the operation – he was a driver. He made people do things – by example. He said, 'If I can't do it, then you can't do it' – but often it was, 'If I can do it, you can do it.' But I only spoke to him once or twice. He certainly wasn't a mixer down on the floor, as far as we NCOs were concerned – he had just a little bit of side. He was number one and he knew it.

Flight Sergeant William Townsend

PILOT, AJ-O

Gibson insisted on keeping schedules and getting in the amount of training we were supposed to do – in fact we got in more training, I think, than we expected. So if he had an idea that he wanted to do something he would do it and he would get other people to do it, so that was his function in life.

Beck Parsons

GROUND CREW

We saw a bit of Guy Gibson – the point about him was that there was little doubt that he was the right man for that job at that time. On a personal level, well, I thought he was a snob, but if I was walking out to dispersal at Scampton and he was in his car, reading a newspaper, on the way out there, and I walked past his car, he'd be out of it – because I hadn't saluted him. I can see him now, and he'd look out. 'Do you know who I am?' But he was the right man for the job. If you look at it all, we got the OK to form a squadron and six weeks later, by 17 May, we'd carried out that raid, which was marvellous.

Leading Aircraftwoman Eileen Strawson (née Albone)

DRIVER FOR GIBSON

I drove Gibson out to his aircraft on the night of the dams raid – he didn't appear to be tense or nervous – but I knew this was the big one.

Eve Moore (later Eve Gibson)

I never really knew him. He kept his innermost thoughts to himself. His first love was the Air Force and he was married to whatever aircraft he happened to be flying at the time. I only came second.

CHAPTER 3

Science and Research

Now that the government and the military had expressed an interest in breaching the Ruhr dams, scientists and inventors had to start racking their brains. The inherent problems were how to deliver a bomb or mine of sufficient explosive power to shatter the massive dams – and how to deliver it absolutely on target. Time after time scientists presented hare-brained ideas to the Air Ministry and the Ministry of Aircraft Production, only to have them rejected as implausible, impracticable or downright ridiculous. It was not surprising therefore, when a project set out by Barnes Wallis, the Assistant Chief Designer in the Aviation Section of Vickers Armstrong, met with similar suspicion. Barnes Wallis was essentially a peace-loving man, but he felt strongly that the right way forward was to destroy the dams, and thereby the Ruhr's industrial power source.

Having already invented the revolutionary geodetic construction used in the Wellington bomber, Barnes Wallis was better qualified than most to put forward a proposal – and he had a reputation for having a brilliant mind, utter dedication and perseverance in achieving his objective. He first proposed a plan for a massive ten-ton 'Earthquake Bomb', to be delivered by a specially designed 'Victory' aircraft, and in March 1941 he distributed a detailed paper, 'A Note on a Method of Attacking the Axis Powers'. Prompted by this, a committee was set up – the Air Attack on Dams (AAD) Advisory Committee. The general consensus was that such a massive bomb would be impossible to deliver – and the military were unwilling to divert resources from the main war effort to a one-off aircraft design.

So, if he was not to have an 'Earthquake Bomb', Wallis needed to find a way of making a smaller charge perform like a really large one – and find a way of delivering it accurately. He applied himself to the task with his customary determination, and his research was greatly advanced by a discovery made by A R Collins, a scientific officer in the concrete section at the Road Research Laboratory, Harmondsworth. Through meticulous experiments Collins discovered that the effectiveness of an underwater explosion could be enormously increased if it were detonated in contact with the target – a piece of intelligence he was quick to share with Barnes Wallis. Larger-scale tests conducted against the disused Nant-y-Gro dam in the Elan Valley near Rhyader, confirmed that a dam the size of the Möhne could be breached with some 6,000 lb of Torpex. Complete with casing, the bomb might weigh between 9,000 and 11,000 lb, but this would be within the capabilities of the Avro Lancaster to carry to the target.

The next question was how to deliver it accurately and get it to detonate at the vital point. Release from low level would assist in aiming accuracy, but if the weapon had to enter the water and sink before exploding, there was no guarantee it would remain in contact with the face of the dam. Nothing could be floated along or sent below the surface, as the water-side of the Möhne was protected by floating booms from which steel nets were suspended.

Gradually an idea began to form and become refined in Wallis's

IWMHU24233.

Barnes Wallis, of Vickers-Armstrong. He was essentially a peace-loving man, but he felt strongly that the right way forward was to destroy the dams.

mind. If he designed a spherical bomb and gave it backspin before release, it might hit the water and ricochet in a series of decreasing bounces to strike the target. In doing so it would overcome the boom and net defences. Furthermore, when it struck the dam and began to sink, the backspin would cause it to cling to the dam wall until it reached the optimum depth, which he calculated to be 30 feet, where it could be detonated by a hydrostatic pistol.

Wallis first experimented with marbles fired across a water-filled tin bath in his garden, and small spheres catapulted across a local lake. To confirm the principle and determine the scientific laws, he transferred his experiments to a ship-testing tank at the National Physical Laboratory (NPL) at Teddington, using spheres of different materials and densities.

Once he had sufficient data, he obtained permission to build a number of wooden and metal spheres and convert a Wellington to carry them. The first airborne drop tests took place in December 1942 over the lagoon enclosed by Chesil Beach. After initial failures, he found that by strengthening the casing and adjusting the height and speed of the aircraft and the rate of spin, he could get the spheres to behave in the manner he had predicted. The Aircraft and Armament Development Advisory Committee continued to support Barnes Wallis's research, and he presented a second paper, 'Air Attack on Dams', to the Air Ministry on 9 January 1943.

The weapon, codenamed the 'Highball', had already gained support from the Navy, who envisaged that it might be carried by an aircraft such as the Mosquito and used to attack shipping (notably the German battleship Tirpitz*). However, Air Marshal Sir Arthur Harris, Commander-in-Chief of Bomber Command, was still not convinced that the weapon was viable. It was only after intervention by the Chief of Air Staff, Air Chief Marshal Sir Charles Portal, that on 26 February 1943 Wallis was given the go-ahead to develop a full-sized version of the bomb to attack dams. On* *the understanding that no more than three of his precious Lancasters would be converted until the weapon was proven, on 17 March, Air Marshal Harris gave instructions for Wing Commander Guy Gibson to form a new squadron of skilled crews and train for its use.*

The race was now on for Wallis. With a mid-May deadline for the operation, there were only eight weeks in which to perfect and manufacture the full-scale weapon, codenamed 'Upkeep', and for the aircrews to train in the precision flying essential for its accurate release.

Norman 'Spud' Boorer

AIRCRAFT DESIGN ENGINEER

Barnes Wallis was a Victorian, and a great 'Empire' man, and the country meant an awful lot to him, so anything he could do which he thought would help the country, that's what he felt he was put on this earth to do.

Barnes Wallis

AIRCRAFT DESIGNER

I thought of what would be an engineer's way of stopping the war, and that would be to cut off the power supply to their great armament factories in the Ruhr. Which involved bombing and destroying the dams, because it takes 100 to 150 tons of water to make one ton of steel and if we rob them of all their water supply, they couldn't produce steel and the war would come to an end.

Early in 1942, I had the idea of a missile which, if dropped on the water a considerable distance upstream of the dam, would reach the dam in a series of ricochets, and after impact against the dam, would sink in close contact with the upstream face of the masonry.

Mary Stopes-Roe

BARNES WALLIS'S DAUGHTER

I remembered the occasion when we gathered round the elderly galvanised washtub, filled to the brim, and 'helped' my father shoot marbles from a large wooden catapult over the surface of the water, measuring the number of hops. 'Playing marbles' became my mother's explanation for his incessant activity in following months, and I thought nothing of it. That was a pleasant game on the sunny terrace, not very different from happy times before the war, by the sea in Dorset on the Isle of Purbeck. For many years, my father organised a family camp, run with military precision, under the downs that run from Swanage to Corfe. When the sea was calm, my father taught us to skim flat stones. Mine went plop, plop and sank. His would slide smoothly with six or seven hops and quietly submerge.

My father was not secretive. I can hear him now, describing to a friend some interesting feature of his work, laughing, 'frightfully secret, my dear fellow'. We children were not forbidden to talk – I never had the impression that these were forbidden topics and so it never occurred to me that they might be worth describing. It was different for my brother Barnes, eighteen months my senior, who was already able to understand designs and their mathematical and engineering significance.

Norman 'Spud' Boorer

AIRCRAFT DESIGN ENGINEER

At the meeting with the Ministry of Aircraft Production, they more or less laughed at him and said his Earthquake bomb was ridiculous – they didn't have an aircraft capable of carrying it. Wallis said he had a drawing of a six-engine aircraft which he'd designed back in Weybridge – and if he were to build this aircraft, it would

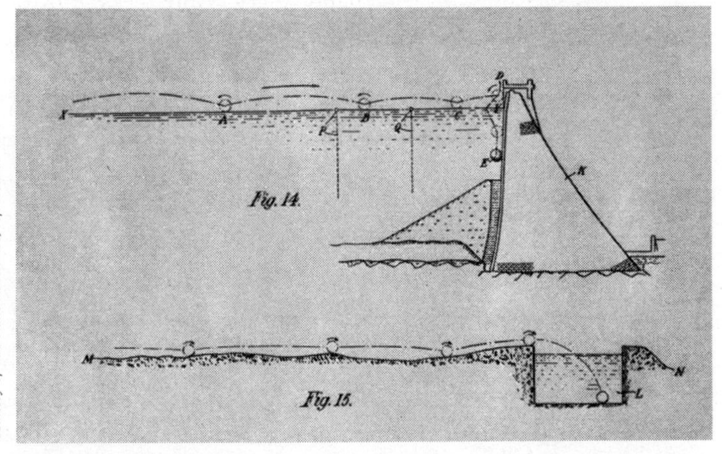

Barnes Wallis's diagrams showing how the bouncing bomb would be deployed over water against a gravity dam (top) and over ground against a canal (below).

carry the bomb. But with all the requirements for labour, this was just not on, so effectively he was told to go back to Weybridge and design his aeroplanes, and stop messing them about.

Barnes Wallis

AIRCRAFT DESIGNER

In a letter to AR Collins Laboratory, written long after the war:
The bouncing bomb was originated solely to meet the requirement so convincingly demonstrated by your experiments that actual contact with the masonry of the dam was essential.

Wallis also wrote, post war:
Just before April 1942, in view of the discouraging results obtained from the experiments authorised by the AAD Committee, I had approached Dr Pye in confidence, telling him that I had an idea which would probably enable a charge to be placed in contact with the dam face and exploded at any required depth, asking him if he would guide the Committee before issuing their report to recommend a final series of experiments with the object of determining the smallest possible charge that would breach the dam when detonated in actual contact with the masonry together with the depth below the surface at which it should explode.

Norman 'Spud' Boorer

AIRCRAFT DESIGN ENGINEER

I could sympathise with the top brass thinking that this was a crazy idea – and at that time in the war, there were many, many crazy ideas being put forward by all sorts of scientists – and most of them were tried and didn't work. This was won eventually by Wallis through his own perseverance and sticking to the job and persuasive powers – and went far enough to prove that it actually could work. Perhaps some of the others might have done if they'd had Wallis running them, and not some other inventors.

Barnes Wallis

AIRCRAFT DESIGNER

In a letter, dated 30 January 1943, to Lord Cherwell, Churchill's senior scientific adviser

This is a report on the effect of destroying the large barrage dams in the Ruhr Valley, with some account of the means of doing it . . . Large scale experiments carried out against similar dams [in fact, just one dam] in Wales have shown that it is possible to destroy the German dams if the attack is made at a time when these are full of water (May or June). It is felt that unless the operations made against the dams are carried out almost simultaneously with the naval operation, preventative measures will make the dam project unworkable and that therefore the development of the large sphere of five tons weight should be given priorities equal to those of the smaller weapon. In the Ruhr district the destruction of the Möhne Dam alone would bring about a serious shortage of water for drinking purposes and industrial supplies. In the Weser district, the destruction of the Eder and Diemel Dams would seriously hamper transport in the Mittland Canal and in the Weser, and would probably lead to an almost immediate cessation of traffic.

A R Collins

SCIENTIFIC OFFICER, ROAD RESEARCH LABORATORY

Of Dr W H Glanville, Director of Road Research Laboratory

His advice represented a turning point in the experimental work because it gave us the confidence to undertake a second test on Nant-y-Gro, a small disused dam in the Elan Valley near Rhyader, with a real hope of success. If this test had failed, and the dam had been severely damaged but not breached, the case for an attack would have been seriously weakened.

Opposition to Wallis's plan

Despite support for Barnes Wallis's project from both Cherwell and Air Marshal Sir Robert Saundby, SASO Bomber Command, Sir Arthur 'Bomber' Harris's response was disparaging.

Sir Arthur 'Bomber' Harris

AOC Bomber Command

This is tripe of the wildest description. There are so many ifs and ands that there is not the smallest chance of it working. I don't believe a word of its supposed ballistics on the surface. At all costs stop them putting aside Lancasters and reducing our bombing effort on this wild goose chase . . . [It is] another Toraplane [flying torpedo] – only madder. The war will be over before it works – and it never will.

To Wallis when Wallis went to show Harris films of tests:
What the hell do you damned inventors want? My boys' lives are too precious to be thrown away by you.

In a later pronouncement, as his aides still supported the project:
I am now prepared to bet that the Highball is just about the maddest proposition as a weapon that we have yet come across – and that is saying something. The job of rotating some 12,000 lbs of material at 500 rpm on an aircraft is in itself fraught with difficulty. The slightest lack of balance will just tear the aircraft to pieces, and in the packing of the explosive, let alone in retaining it packed in balance during rotation, are obvious technical difficulties . . .

I am prepared to bet my shirt (a) that the weapon itself cannot be passed as a prototype inside six months; (b) that its ballistics will in no way resemble those claimed for it; (c) that it will be impossible to keep such a weapon in adequate balance either when rotating it prior to release or at all in storage and (d) that it will not work, when we have got it . . . Finally, we have made attempt after attempt to pull off successfully low attacks with heavy bombers. They have been, almost without exception, costly failures. While nobody would object to the Highball enthusiasts being given one aeroplane and being told to go away and play while we get on with the wars, I hope you will do your utmost to keep these mistaken enthusiasts within the bounds of reason and certainly to prevent them from setting aside any number of our precious Lancasters for immediate modification.

Gradually, Harris was won over, influenced mainly by Sir Robert Saundby.

Norman 'Spud' Boorer

AIRCRAFT DESIGN ENGINEER

Barnes Wallis was summoned to Whitehall to meet with the Chief of Air Staff. He said afterwards that he came out of that office and he really felt sick, because he realised at last that his bluff had been called, and he'd got to get on and do what, for years, he'd been saying he could do.

Sir Henry Tizard

SCIENTIFIC ADVISER TO THE AIR MINISTRY

I should myself be inclined to advise that Wallis be instructed straight away to submit an opinion as to whether a bouncing bomb of this size [8,000 lb] could be fitted to a Stirling or a Lancaster.

Sir Arthur 'Bomber' Harris

In letter to Sir Roy Dobson of A.V. Roe & Co. (Avro):
The Lancaster surpassed all other types of heavy bomber. Not only could it take heavier bomb-loads, not only was it easier to handle, not only were there fewer accidents with this than with any other type throughout the war, the casualty rate was also considerably below other types.

I used the Lancaster alone for those attacks which involved the deepest penetration into Germany and were, consequently, the most dangerous. I would say this to those who placed that shining sword in our hands: 'Without your genius and efforts we could not have prevailed, for I believe the Lancaster was the greatest single factor in winning the war.'

Wing Commander Guy Gibson

At Grantham I saw the SASO, who said, 'I'm sending you on a journey down south to meet a scientist who is working on your project. He's going to show you nearly everything, but remember, only the AOC and myself and five others know anything about these matters, and you'll be the eighth. I can't stress too much the need for secrecy. It's absolutely vital.'

We went on, down the winding Great North Road, meeting very little traffic except vast army convoys going north, past London into the southern country and up to an old country railway station. There I caught a train. My destination was so secret that not even my driver was allowed to have any idea of where I was going. Half an hour later, I was met by a tall man whom I shall call Mutt [this was Mutt Summers, Vickers' chief test pilot]. He was the senior test pilot for a very well-known aircraft firm, and had himself been responsible for testing the prototypes of more than one of our most successful bombers.

We drove in his little Fiat car quite a way, without saying a word. I don't think he expected to see such a young fellow, and I didn't expect to see a civilian, and I think we were both wondering what we were doing there, anyway. At last he pulled up at an old country house. Here our passes were checked and rechecked and I had to pull out my special buff-coloured pass (numbered 7), which the SASO had given me earlier in the day. Then a couple of tough policemen gave us both the once over and we entered this queer place. We went down a long, dimly lit corridor, down dark stairs, further and further into the earth. Mutt seemed to know the way, and at last we came to a large iron door. There were two more guards here, and once more we went through the same procedure. They were certainly being careful. Then one of them opened the iron door and we went into a sort of laboratory. It was bright inside, much brighter than the dim corridors, and I blinked to try and get accustomed to the light.

Then I met a man whom I'm not going to try to describe in detail, for I know he wouldn't like it. But I'm going to call him Jeff [this was Barnes Wallis], which of course is not his real name. He was a scientist, and very clever aircraft designer as well. Jeff was neither young nor old, but just a quiet, earnest man who worked very hard. He was one of the real back-room boys of whom little can be told until after the war, and even then I'm not sure their full story will be told. He looked around carefully before saying anything, then abruptly, but benignly, over his thick spectacles, 'I'm glad you've come; I don't suppose you know what for.'

'No idea, I'm afraid. SASO said you would tell me nearly everything – whatever that means.'

He raised his eyebrows. 'Do you mean to say you don't know the target?' he asked.

'Not the faintest idea.'

'That makes it awkward – very awkward.'

'But the SASO said –'

'I know, but only very few people know, and no-one can be told unless his name is on this list.' He waved a list in front of me. I could see there weren't many names on it.

'This is damned silly,' said Mutt.

'I know, but it can't be helped – but I'll tell you as much as I dare. I hope the AOC will tell you the rest when you get back.'

I said that I thought this would be all right and waited, very curiously, for him to go on. Then he said, 'There are certain objectives in enemy territory which are very vulnerable to air attack, and which are themselves important military objectives. However, these need a vast amount of explosive placed very accurately to shift them or blow them out – you know what I mean – viaducts, submarine pens, big ships and so on. I have had my eye on such things for a long time, but always the problem has been too great – much too great.

'First of all, there wasn't an aeroplane with a high enough performance to carry the required load at the required speed. Then along came the Lancaster bomber, and this problem was solved. The next one was the explosive itself. It would have to take the form of either a very large bomb or a very large mine. But if it was to be dropped accurately enough to do its job, it would have to be placed within a few yards of the right spot. There are three snags to this. If bombing is to be as accurate as that, then the attack will have to be at low level, which means below 300 feet. But with these great big bombs there's always the danger that they may explode on impact from this height – and you know what that means. And, if they're dropped above that height, then accuracy diminishes and the job can't be done. It's a sort of vicious circle. The other two snags are, of course, the danger of flak at that level and balloons, and the difficulty of flying over water at low level.'

'Over water?' I said.

'Yes, over water at night or early morning, when the water will be as flat as a millpond backed up with a lot of haze or fog all around.'

I began thinking of possible targets. *Tirpitz*, U-boat pens? No future in this – but Jeff was still talking.

'For a month or two now, we have had the go-ahead order from the War Cabinet to try and overcome these difficulties. So we've been working hard, Mutt and I, on a certain theory of mine. I discovered the idea quite simply, but you won't want to know about that. Come, I'll show you.'

The lights went out in the lab and a small screen lit up with a flickering motion picture. The title was simple. It read, 'Most Secret Trial Number One'. Then an aeroplane came into view, diving very fast towards the sea in a sort of estuary. When it got to about 200 feet, it levelled out and a huge cylindrical object fell from it rather slowly towards the water. I was amazed. I expected to see the aircraft blown sky high. But when it hit the water there was a great splash, and then – it worked. That's all I can say to describe it – just that it worked . . .

When the screen became white and the lights went up, Jeff said, 'That's my special mine to overcome our difficulties, and it does work. But I'm afraid it's only one quarter the size of the real thing required to do the job. When we get to the big fellows, I think we're going to run into a lot of difficulties.'

'Has any been made yet?' I interrupted.

'No, not yet; the first will be ready in about a week's time with a modified Lancaster to carry it. Avro's are doing a great rush job to get the special fittings put on; I believe they're working twenty-four hours a day. Now what I want to know from you is this. Can you fly to the limits I want? These are roughly a speed of 240 miles an hour at 150 feet above smooth water, having pulled out of a dive from 2,000 feet, and then be able to drop a bomb accurately within a few yards.'

I said I thought it was a bit difficult, but worth trying. I would let him know as soon as possible. Then I was on my way. Out of this strange house and into the open air again. Mutt drove me to the station and four hours later I was back at Scampton.

George Robert Edwards

VICKERS AIRCRAFT ENGINEER

It was difficult from the technical point of view of getting the bombs into the parts of the dam where they were most vulnerable. It took a hell of a lot of doing.

Nobody had ever done anything like it before, so there was a lot of trial and error. We used the water tanks at the NPL – but that was a hell of a long way from doing it with a Lancaster. In the early days, I was doing the experimental trials and Wallis and others were of the opinion that the bomb needed topspin on it. I was quite sure that it didn't – that it needed backspin because as a bit of a bowler I knew enough about the difference between topspin and backspin, especially as far as water went. I had some difficulty in convincing the powers that be that that was how it was, and I finally convinced them by getting a catapult built with an arrangement so that I could fire one with a bit of backspin and another with no spin at all, and then fire another with topspin. And to everybody else's surprise, the one with topspin screwed itself in, the one with no spin hardly bounced at all, but the one with backspin skittered along the lake ten or a dozen times. That was at a lake in Cobham.

I knew about this lake, because I'd been running an experimental shop on the edge of it. I went and saw the lady who owned it and said I wanted to borrow her lake. I had to explain that it was part of the operation to defeat Hitler, and she said, 'In that case, my boy, you take the lake and you do anything you like with it.' So off I went with the lady's blessing.

Wing Commander Guy Gibson

At Scampton the AOC sent for me, and as I entered his office, I noticed three large packing cases on the floor; he handed me a screwdriver. 'These are models of your targets,' he said. 'Now I'm not going to tell you where they are, nor am I going to tell you what they are, although of course, you will probably realise as soon as you see them. Jeff has rung me up and told me that you won't be able to train your squadron unless you know, and so I am leaving it to him to tell you all the details. However, don't forget that you have got to be the only man in the squadron who can possibly know the target until the day before the attack is made.'

The three models were perfect in every detail down to the smallest tree, and my first feeling was, 'Thank God it's not the *Tirpitz*.' This was something I did not expect. There were models of three dams, and very large dams at that.

It was later, in his country house base that 'Jeff' opened his file and began: 'The dams which you saw in model form in the AOC's office are great barrage dams in the Ruhr Valley. The weapon you saw in my laboratory the other day is known as Downwood, and it's my idea that by really accurate use of this weapon we shall be able to knock down the concrete walls of these dams.'

'But surely a smaller bomb would do that easily?'

'No,' Jeff laughed. 'A lot of people think that – they think dams are just curved structures that hold back water by their shape, much the same as the arch of a bridge. There are dams like that, but they are known as vault dams. These barrage dams are known as gravity dams, and hold back the water by their weight. As these are 140 feet thick of solid concrete and masonry, and 150 feet high, you can see that there is a colossal amount of masonry to shift.'

He went on 'The Germans are very proud of this dam. In fact, it is rather beautifully built, Gothic and all that. It is some 850 yards long and 140 feet thick, and high as it is thick, and the lake

IWM MH3781 *and* IWMMH27710.

Scale models of the Sorpe (top) and Eder (bottom) Dams, constructed from aerial reconnaissance photographs.

it holds back is about twelve miles long, holding 140 million tons of water. At the same time they built another dam nearby, called the Sorpe. This is very much smaller and is of earthen construction, which, if you know what I mean, consists of a sloping bank of earth 600 feet long on either side, holding up a watertight concrete core. Between them they hold back about 75 per cent of the total water available in the Ruhr Valley. If they were to be breached, the consequent shortage of water for both drinking and industrial purposes might be disastrous. Of course, the damage done by floods if they were breached quickly would result in more damage to everything than has ever happened in this war.

'There is a third dam I must tell you about, and this is the Eder. This dam is some sixty miles away. It was built in 1914 primarily to prevent winter flooding of agricultural land and to assist in improving navigability of the lower Weser River. The dam also provides some water for the Mittelland Canal, which, as you know, is one of the main canals in Germany, running as it does on the main route from the Ruhr to Berlin. But unlike the Möhne Dam, the Eder has no water-supply functions. However, it supplies a lot of hydroelectric power. This one's a little larger than the Möhne Dam and lies in a valley forty miles from Kassel. And it holds back 202 million tons of water – but to breach these things is an entirely different matter.'

Jeff explained, 'We built out in the garden a dam some 200 feet across. This dam, I might tell you, was brick for brick in strength the same as the Möhne Dam. The lake was filled with water, and by manipulation of certain charges, we intended to try out there the theories we had evolved from the smaller models.

'After a lot of trouble, we got a charge in the right place, which we worked out a Lancaster on the same scale could carry, and the dam wall cracked. It cracked round about its base, and after a few more charges had been detonated, the wall moved over backwards and water ran down into the garden. But this was not enough –

we had to still make tests on a full-size model. At that time we heard that a certain County Council in the Midlands had just built a new dam to supply their town with water. We heard about it and wrote to them and asked them if we could knock down their old dam, so that the water would run into their newly built one. After a lot of quibbling they agreed to let us do this – and once more after much trouble, we succeeded in knocking it down.

'My smaller-scale weapons have worked, but we haven't had a chance to try the big ones yet – they aren't ready. But I think they will be in a few days' time, and so the trials are set for the 16th. If the big ones work, Avro's will have to modify twenty-five Lancasters to carry them. This is quite a big job, because these mines weigh a lot and are about 11 feet in diameter. Then there's the time factor. The ordnance factory will have to construct the weapons, you'll have to plan a special method of attack, and all this will have to take place within a month.'

'Why the urgency?' I asked.

'Because dams can only be attacked when they are full of water. Every day photo planes are going out to take pictures of these dams, and we are watching the water rising. At the moment it is 12 feet from the top, but we can only attack them when the water is 4 feet. This ensures that the maximum amount of water is in the dam and at the same time makes certain there is 4 feet of free board on the lip against which you have to throw your mines. That is why you have to be so accurate. I have calculated that the water level will be suitable during the week of the 13th to the 19th of May – that is, in about six weeks' time. This, as it happens, is a moon period – I think you will have to do it at night or dawn – you couldn't get into the Ruhr by day, could you?'

'God, no.'

'Then the moon will be useful, but if you want more light, you'll have to do it at dawn – but that's up to you. Your projectiles will have to fall so that they sink into the water actually touching the

Courtesy of The National Archive, ref. AVIA10-369.

Barnes Wallis and his team carried out large-scale tests on a small dam at Nant-y-Gro in the Elan Valley, near Rhayader.

dam wall itself about 40 feet down; if they are not touching, it'll be useless. Then, when the mine explodes by a hydrostatic fuse, I have calculated that crack will appear just as it did in the models. By placing more mines in exactly the same spot, you will be able to shift this wall backwards until it rolls over, helped, of course, by the water pressure. The Sorpe Dam requires rather a different technique, but we will go into that later.'

Patricia Reid

Assistant Section Officer – Intelligence branch, photographic interpretation

There were three different grades of interpretation – not according to how good the people were at doing it, but according to the usefulness of the information. There was the first interpretation – first phase – done on the RAF stations where the aircraft had flown to take the photos. The moment they came back with their pictures they were rushed to the first phase people who would do an immediate evaluation of them and put in a quick report, because it might be that they contained something that needed immediate action. The photos were then sent on to the second phase, where things were done in a great deal more depth – again reports from that would go to where the information was needed. Then there was the third phase, where people would have these photos and work on them for a long time, in even greater depth, with a great deal of other research material. The people who had done best on the course were sent to one or other of these.

I passed, but didn't do particularly well on it, but it worked out very well for me, because I was posted to London to the Air Ministry, where there was a branch called ADIPH. There was the Intelligence branch, and as one of its subsidiaries it had the assistant chief of the Air Staff director of Intelligence Photography – and they had a committee which sat and coordinated requests coming in from organisations wanting the photographs to be taken. We would sort it out and pass it on with instructions as to what sort of photographs had to be taken, and at what time – and how, because they used to take low-level obliques if it was somewhere that the planes could get in very low. More often it had to be very high-level vertical photographs.

It was fascinating, balancing the requests with the instructions to the pilots who would do it. It was quite a small department of eight or ten people – I was in a very junior capacity, working with more senior officers but it was a very interesting job.

Then we started getting requests through for photographs to be taken of the Möhne, Eder and Sorpe dams in Germany. They used to send out special forms to the RAF stations where the planes were, and the pilots were going to fly and take the photographs, and a lot of work went into getting the requirement sorted out and exactly tabulated as to how it was to be done if possible to do it. Everything was done on scrambler telephones, special green phones, and the usual thing was when someone phoned you up, they'd say, 'Shall we scramble?' and that meant you pressed the button and it safeguarded the information being passed. It was a very secret planning arrangement. Then, when we got the photos back, which everyone was longing to see, because Guy Gibson's plan had been much discussed at very high level – Churchill downwards – and they were all very keen to know if this marvellous plan for Barnes Wallis's bouncing bombs would work or not. It was terribly important that the photos should show how high the embankments were – the walls – so they could work out the height of the dam and the possible flow of the river. The photos had to establish so much information beforehand, for there to be any hope of the pilots being able to achieve what was wanted.

Everyone was longing to see these photos – and the eventual interpretation of them. On one very exciting day, I was told by my boss, 'Look, there isn't a dispatch rider handy – we'll ring up and

Aerial photograph of the Möhne Dam, used for determining the water level and constructing the briefing model.

get a taxi. Take this – it's sealed – go round to the Foreign Office in King Charles Street. Get the taxi driver to step on it and hand these in.'

I handled this parcel with awe, because I knew how much this was wanted, and that it was very important.

Wing Commander Guy Gibson

Bob Hay and I went to Reculver, where we had been told the tide would be just right for the trials at seven o'clock next morning. We were told that after each mine had been dropped, the scientists wanted to examine them to see how they had stood up to the shock of hitting the water at that very high speed. The idea was to drop one when the tide was on the ebb so that when it went out, we could walk down on the beach and examine it thoroughly without the difficulty of recovering it from the sea.

Early next morning we stood together, Bob, Professor Jeff and I, on the range facing the sea, waiting. It was one of those white mornings with the cloud spread evenly over the whole sky; there was hardly a ripple on the water; it was cold and rather grim and our collars were turned up. Jeff looked at his watch and said, 'Shorty should be here any minute now.' Shorty was one of Mutt's test pilots – he was five feet six inches high and rejoiced in the name of Longbottom.

Behind us, round the barbed wire which surrounded the range, special policemen patrolled, their job to keep away strangers. No-one else was meant to see these trials.

Then out of the sun came the two Lancasters. They were both in fine pitch and making a very hearty noise for that time of the morning. They flew in formation; one the motion-picture aircraft, the other with one of our mines slung underneath. Inside the gaping bomb doors we could see it quite clearly, painted black and white, looking large enough even against the massive black Lancaster itself.

Down came the Lanc with Shorty Longbottom at the controls, to about 150 feet, travelling at something like 270. We saw him level out for his run, then climb a bit to get his exact height over the calm water. We saw him tense at the controls, getting his horizon level on the cliffs further on. Jeff stood beside Bob, crouching like a cat. The movie camera began turning. I picked up a pair of binoculars. Then the mine fell out quite slowly. It seemed to hang in the air for a long time before it hit the water with a terrific splash and a dull thud. In a minute we would know whether or not Jeff's calculations had been right or wrong, but for the moment there was nothing except that mighty wall of water which reached up to the aircraft's tail as if to grab it. Then it all subsided and we knew. The great mine could be seen taking its last dive, smashed into six different fragments . . .

In our secret hangar that afternoon there was feverish activity as the mine was strengthened by steel plates so that it wouldn't break when it hit the water. Small men with glasses toiled side by side with sweating airmen, none of whom knew the use for which the thing was intended. They had no meals and worked through an alert without pause. At five o'clock that afternoon they straightened their backs. It was ready.

In the late afternoon we again stood waiting . . . Then over they came again; once more the suspense and once more the mighty splash. Then Mutt, who was flying this time, banked steeply away to have a look. But we weren't looking at him – only at the bits of our smashed-up weapon which were hurtling round the sea like flying-fish. Then the foaming water settled down again and there was a long silence. Jeff suddenly said, 'Oh my God,' and I thought he was going to have a fit – but he soon calmed down, for he was not temperamental, and as we trudged along the shingle to the car park he began planning his next move. Here was a man who would not be beaten.

Barnes Wallis

To make the device work, I had to ask Guy Gibson to come down to a height of 60 feet – but one felt in doing that that one was endangering those men's lives, simply to make an idea work – and that is misery, if you want to know what it's like.

Norman 'Spud' Boorer

AIRCRAFT DESIGN ENGINEER

The net result of the first test of the full size bomb on 13 April was that it didn't work. The first one was dropped from a Lancaster at 250 feet, and dug straight in – nothing happened.

Wing Commander Guy Gibson

One after another the experiments with the mine failed. The mines were not working. I can remember clearly the sight of that cold, quiet figure standing sometimes alone, sometimes with two or three others, on the brink of the water, looking up at the great Lancasters. There was a tenseness about the way in which he stood, with his legs apart and his chin thrust out and a fearful expectancy about everything.

For many days they worked and flew while he modified and experimented and the boys trained and trained, and I watched and watched, But every time it ended in the same thing – failure.

One day Shorty came up to collect me in a Mosquito. 'I have got to take you to Brooklands,' he said. 'Jeff's there, and he wants to have an urgent conference with you.' I needn't say how tired Jeff seemed when I went in. He said to me, 'The whole thing is going to be a failure unless we can jiggle around with our heights and speed.'

'What do you mean?'

IWM FLM2360, FLM2361, FLM2362, FLM2363.

A 617 Squadron Lancaster releases an inert Upkeep at Reculver. 'A number of us were given the opportunity to go down and drop a dummy – to get the feel of how the aircraft would react when the 9,000 lb of revolving steel was released.' Flight Lieutenant David Shannon.

'Simply this. From the slow-motion movie cameras which have taken pictures of this thing dropping, I have found a few facts. I have drawn this graph here to illustrate what I mean. It's all a combination of speed and height. You see here that they will work and won't break up if we drop them from 150 feet at a certain speed. On the other hand, if we drop them from 40 feet at this speed they will also work. The best height to suit your aircraft is here at the 60-foot level at 232 miles an hour. But that is very low, and that's why I have asked you to come down. Can you fly at 60 feet above the water? If you can't, the whole thing will have to be called off.

George Robert Edwards

VICKERS AIRCRAFT ENGINEER

We went down to Reculver for tests, and we were using Lancasters then. We had a lot of people to convince that it was going to work. We had a bit of luck one day there, in that the spherical shape was held with steel bands, and they broke. The metal drum that was inside was left spinning on its own. It travelled along quite nicely as a cylinder, and the subsequent bombs we used were cylinders.

Beck Parsons

GROUND CREW

When the squadron was formed, I was part of a maintenance crew of about six people and they were based in the hangars. They weren't allocated to one aircraft; they were overseeing all the aircraft. My task was with the electrics and radio and that area – plus bomb gear and all the general electrics that all the day-to-day people wouldn't be involved in unless they were in trouble.

The Lancasters weren't normal – the major difference was that the bomb bay was cut away and there were two arms fitted to the side of the fuselage which were sprung – and these two arms had a disc on at the end, and the discs clamped on to the bomb underneath the aircraft and clamped into the rings at the side of the bomb – one of the discs was driven by a little hydraulic motor up in the bomb bay, which was rotated at a rate in the air of 500 rpm – it moved the bomb and when you consider that that bomb weighed 9,500 lb, it was quite a feat.

They cut the mid-upper turret out, because they were carrying more weight than normal – 10,000 lb extra. There were other things that changed, such as the lights underneath the aircraft. Barnes Wallis had said that there were certain parameters that you'd got to observe to get this to work, and he started with the idea that it would work at 150 feet, which meant that you had to bring the aircraft out of a hangar, get a Coles crane behind them and lift them up into a flying position, and you'd got these two Aldis lamps – one forward, one half-way down the fuselage, set at certain angles. And the angles were worked out in the first instance so if you set them at 150 feet, the theory was that when they met that was your height. But at that height the bomb broke up, so he realised he was wrong and he reduced it to 100 feet and had another go. And still the bombs were breaking up and not spinning the way they should. Eventually he decided on 60 feet, which was very, very low – which meant that if you were flying in to a target dam, at 60 feet, the only way to find if it was 60 feet was if these two lights met at a point and became virtually one. Then it was 60 feet and you had to maintain that until you got to the target.

John Elliott

NCO GROUND CREW

We had a couple of meetings where we were told to be quiet about the thing and not to talk to people outside or to write to anybody about what sort of training was being done. When the operational

Courtesy of The National Archive, ref. AVIA53-627.

Patent diagrams showing, above, side view and, below, front view of the mechanism that held and spun the Upkeep bomb.

aircraft arrived we realised that something really special was going on because they were modified. We'd never seen a Lancaster like it before.

Well, they had no bomb bay. Instead they had two great arms on either side of the belly of the aircraft and when they eventually brought these 'bombs', they winched them up between these two arms, which were then closed together. There were two things like pads which fitted into either side of this – like the front of a steam-roller. These arms were eventually winched together so that this thing was suspended between them and there was a drive on one side of it to spin it round. We realised there was going to be something a bit spectacular. While they were training for this operation it was really hard work – some of the hardest work I ever did in the air force.

It was such a peculiar set-up and a peculiar looking weapon – and the fact that it was going to be rotated. We knew it was going to have to be done at a very low level as most of their training was low flying and we eventually got to know that they needed to fly at 60 feet. These aircraft, to arrive at this 60 feet, depended on two spotlights, one forward and one aft, and the angle of these lights was adjusted so that at 60 feet the spots came together. Some clever mathematician got it worked out so that they could adjust these things in a hangar.

A lot of weight had to be put in the tail end to hold the thing down in flying position in the hanger, but with it in flying position in the hangar they worked out how far apart the spots had got to be for them to meet when the aircraft was 60 feet off the ground – a bit of clever geometry I think.

The only different thing about the engines was that we had two extra hydraulic pumps on them and the hydraulic system of the aircraft was modified – but there was additional hydraulic piping installed and bits of hydraulic gear put on. I think they were hydraulically driven and that's the reason we had all this extra hydraulic equipment fitted. That was a bit of a problem for some time – we used to get quite a lot of air in the hydraulic system, and it just wouldn't perform properly. But this thing was driven, as far as I ever knew, by a hydraulic motor. I think they used to call them swash-plate motors. The motor was driven by hydraulic pressure, and they had to spin this to do 500 revs per minute backwards. I realised afterwards that it was rotating backwards so that when it was dropped it didn't sink like a stone. Instead it bounced along the surface, so that it would eventually come up against the dam before it sank. I believe there were pressure fuses in the mine, so that the water pressure at a certain depth would set it off.

Then when the raid took place, we felt we had a proprietary interest in it – although all we'd done was be the channel between the organisers and the doing of it.

Sandy Bonser

AIR FRAME MECHANIC

When we worked on the oleo legs, Tiny Hucksley, who was about six foot eight inches tall, built like a tank and very strong, held the legs in position, while one of the lads and I stood on a gantry and put in the retaining bolts – which was no mean task, as the legs were very heavy.

The arms either side of the bomb had to swing apart when the bomb was released, so this was a problem to get hinges fitted to the main spar to the side of the fuselage. For the arms to work they had to be fitted with hydraulic rams, then another modification was needed to get the bomb spinning. This was a headache. They wanted a motor strong enough and small enough to fit just rear of the centre section, where the old bomb doors finished. They managed to get a motor from a submarine – this worked perfectly.

Kenneth Lucas

GROUND CREW

I spoke with Barnes Wallis – he came when we were fitting the arms that carried the bomb, and he had a look round. I had a few words with him. He was certainly a very clever man, but we didn't know what it was all about – we were mystified at the time. We just did what we were told.

Sergeant Dudley Heal

NAVIGATOR, AJ-F

All the time, we didn't know that they were still working on the bomb, and very close to the operation we were suddenly told that it was to be dropped at 60 feet – apparently that was the best height for that bomb to achieve its purpose. So we all went to the bar and had a drink. Sixty feet! I ask you!

John Elliott

NCO GROUND CREW

Barnes Wallis seemed a typical absent-minded professor – always thinking about something that was going to happen or something that he was going to do. I met him at a dinner after the raid. Always got something on his mind, always thinking of something a bit further on than he had already got. 'What's the next thing that I've got to do?' Or snags, 'How do I get over this?' And he certainly did get over it. Every snag he came up against, he got over.

The people who made the hydraulic motor affair that was going to drive this bomb came down; we got quite a lot of help from them. It was a question of getting your hydraulic system completely free of air, then once you could do that the mine would rotate properly at the speed that they wanted.

We had to bleed the air off – there were various points along the system where you could bleed air from, using spanners and bleed screws. This was the responsibility of the airframe people and we used to see them working so hard at it. I used to get a bit of the backlash because every time they wanted to test the thing, I had to get in there and start and run the engines for them until they got it right.

Sergeant Ray Grayston

FLIGHT ENGINEER, AJ-N

I knew who Gibson was, and I saw him on the base, but I had nothing to do with him. He didn't recognise NCOs at all. But Barnes Wallis – a great guy. I knew him because I worked for Hawker Siddley and he worked for the enemy up the road, Vickers.

Beck Parsons

GROUND CREW

We saw Barnes Wallis at Scampton – but I was very much in awe of him. I was tied up in fitting all the lights and the electrics, just as he wanted them, and I was rather in awe of him then – but by the end of the war I knew how clever he was. My first impression was always good.

CHAPTER 4

Training at Scampton

By the end of March 1943 Gibson had got his team assembled at Scampton. Under instructions that their work was to be conducted in the strictest secrecy, Gibson explained the nature of their forthcoming training. The one-off mission they were to carry out would entail low-level flying by night over water – so the crews duly began a period of seven weeks' intensive training, during which those who had not arrived as part of a crew would have to bond with their new teams and learn to work together. Airmen who did not match up to Gibson's exacting standards were sent back to their squadrons while 617's elite crews honed their performance to ever more exacting criteria.

Flight Sergeant George Chalmers

WIRELESS OPERATOR, AJ-O

I arrived at Scampton with little idea about what was going on, because the first thing I had to do was spend an hour and a half reading the security orders. Then I was told to report to the squadron

IWM ATP011384B.

Type 464 (provisioning) Lancaster. This aircraft was delivered to Scampton as a spare, but was flown on the raid by Flight Lieutenant McCarthy and crew.

commander down on the airfield. When I got there I discovered that we had no aircraft and that the hangar – with nothing in it – was guarded both ends by police. When I finally did meet the squadron adjutant he informed me that they were putting a squadron together – and that was the first I knew about it. More than that I wasn't told. I just sat back and waited for events to happen. Then we got the aircraft and went into intensive training for low flying.

John Elliott

NCO GROUND CREW

G-George was no more trouble than any other Lancaster. When 617 was formed, each crew got to have two aircraft. In the first instance they were standard Lancasters, which they used for all the training until we got our operational aircraft. We thought in the first instance when we all got posted to Scampton to form this squadron, that Gibson was just going to form another ordinary squadron within the group. We didn't think any more of it – there's a tremendous amount of work in getting another squadron together. You've got all your ground equipment to get organised, all your flight work to get done, and then when you want to start flying, things get really busy.

Sergeant Ray Grayston

FLIGHT ENGINEER, AJ-N

I thought the Lancaster was a marvellous machine – it was the best of anything flying at that time.

Low-level flying

No-one on the squadron had any idea why they were training at such low level – the reason would only become apparent when they were briefed.

Flight Sergeant George Chalmers

Wireless operator, AJ-O

None of them, as far as I could make out, had done any low-level flying before. For the pilots this was something new to learn, and in the early stages some of the crew members were holding their breath till they got back down again. What was surprising was that we began to enjoy it.

Flight Sergeant Grant McDonald

Rear gunner, AJ-F

We had no idea what we were training for on 617. Low-level flying was a bit worrying and quite a bit different from what we were used to. At that time Bomber Command had been attacking targets from higher and higher. We had all been stacked up at 20,000 feet and above. Suddenly it was low-level and quite alarming in the rear turret, watching the ground go by so quickly. You heard a lot about people flying under high-tension wires and so on. Some aircraft suffered damage through hitting the tops of trees.

Flight Lieutenant Les Munro

Pilot, AJ-W

Once we started the low flying in training, it was exhilarating. It was an enjoyable change from the high-level trips we had been used to. It would depend on the ability of the particular pilot, but I enjoyed low flying. Most of the crews did, because normally it was a court-martial offence if you did it unauthorised.

Flight Sergeant Ken Brown

Pilot, AJ-F

We started our low flying and you've heard various stories about how we started at 60 feet. It really wasn't so. We started our low flying cross-countries over England at about 200 feet. That lasted about three days. Then we were down to 150 feet.

Flying Officer Harold Hobday

Navigator, AJ-N

On our very first briefing with Gibson, he just said that we'd got to fly low over water, and get down as low as we could all the time, and navigate first of all in daylight. It's very difficult to navigate at about 60 to 100 feet. Then we'd got to do simulated night flying, which entailed having blue screens around the aircraft so that it looked like a moonlit night – so you couldn't see very well. We did this in the same places we flew over in daylight – it just made everything more difficult to see – as we would find on the raid itself. When doing simulated night flying it was very difficult to navigate with maps, but we got used to it in the end. It got us used to the real thing. Towards the end of training we did fly by night with as much moonlight as one could get – but it was still a bit dicey. We did a lot of flying like that – and a lot of low-level bombing practice.

Flight Lieutenant David Shannon

Pilot, AJ-L

To start with, Gibson just said that our brief was to practise formation flying in threes as low as we could get. It was left to us what height we considered to be safe. Then he said that he wanted us to become proficient down to 150 feet. The altimeters on the Lancasters were pretty dicey then, so this needed practice. We were flying all over the country at 150 feet; our navigation officers would just map out a route that they thought was suitable.

Once we'd perfected that, and could navigate all round the country, we started flying at night in moonlight periods. Well, there's not an awful lot of moonlight available in this country, even in the summer, so we did a lot of very early morning and dusk flying to try to simulate the light. Even then, we weren't getting sufficient practice, so I think it was the Americans who came up with a way of simulating moonlight. The cockpit of the aircraft was covered in blue Perspex and we were to wear yellow goggles – complementary colours – and fly during daylight in these adapted aircraft. That was as near as we could get to moonlight flying. We all did two or three cross-countries under these conditions to get in the practice we wanted.

This was quite successful – but a bit hairy to start with, especially taking off. On a bright day one tended to want to take off the goggles and see what was going on, but we very soon got accustomed to it.

We'd already done a certain amount of low-level flying in my first tour of operations on daylight raids – at that time all daylight flying was done at low level to avoid any fighters, because if you kept down on the deck, the chances of fighter attacks were greatly lessened. But I don't think anybody in Bomber Command had done continual low-level flying to the extent we were expected to do to train and reach our level of perfection.

Flight Sergeant Ken Brown

PILOT, AJ-F

I did a cross-country one day, and I came across a new aerodrome that was being built with an awful lot of people around it. There I was headed straight for the hangar and I thought I'd better pull up, as there's no point in trying to go through it.

The Royal Observer Corps kept track of us all the time so Guy got our altitudes no matter where we were and had a report on them the next morning. So at briefing the next day he said, 'Brown, what were you doing going over the hangar?'

'I thought it was a good idea.' I said.

'Two hundred feet!' he said. 'Hardly, you'll do that one again.'

It wasn't a bad cross-country anyhow, so I did it the next day.

When I came to the hangar – same thing. All these men were working on top of the hangar and this side of it and so forth. So I put the aircraft down on grass level and then came over the top of the hanger and there were people sliding off it and running in all directions. So next day at briefing, Gibson looked in my direction and said, 'Brown, I said low, but not that low.'

Flight Sergeant Leonard Sumpter

BOMB-AIMER, AJ-L

We used several dams for our practice – the Derwent Dam was quite suitable because it was in a valley. We didn't know then what we were going to attack – or even why we were doing all this flying over water, but this was very like the Möhne Dam.

On a typical training day you'd get up in the morning and pick up the other NCOs in the crew – Brian Jagger and Bob Henderson and myself. All other others were commissioned ranks. Shannon would let us know the day before what time to be at the flights next day.

You'd go down into the crew room and meet together, and the navigation officer had a number of set cross-country flights which we had to do. He'd say, 'Number 2 today', or 'Number 5'. Then you'd go into the navigation room where there was a table, and the navigator would do his pre-flight planning. He'd draw all your tracks in – where you were going – and sometimes, by the second or third week, you got some maps with all the numbers on. You had your set of maps for each numbered flight, and you'd wait until the navigator had done his pre-flight planning, taking into account the take-off time, the winds and ground speeds.

Then you would go to the parachute room and sort out your harness and parachute – your helmet you'd take with you. You'd just go out to the plane and take off on your own – no set time – and you go out and come back again. After that you'd go back to the flights, hand all your parachute kit in and have a little chat over whether things went right or wrong – or how the system could be improved. Then you just went off to your respective messes – officers to theirs and sergeants to their own.

Aircraftwoman Morfydd Gronland

The Sergeants' Mess doors would burst open and the aircrew would swarm in, shouting boisterously. We young WAAFs had to endure a barrage of good-humoured banter: 'How's your sex-life?' 'I dreamed about you last night.' 'Please serve us in the nude.' Then someone would ask, 'What's the collective noun for WAAFs?' And a chorus would answer, 'A mattress.' But we took it all in good part because we knew the great strain they were under and the dangers they would soon face.

Flight Sergeant Ken Brown

PILOT, AJ-F

We were entirely dependent on the efficiency of the ground crew, and they have to take a great deal of credit for keeping the planes in the state that they could fly to a target hundreds of miles away and come back, and they've got to take the credit. I think insufficient credit has been given to ground crew and the part they played in Bomber Command operations, and they had a major influence. I was always very grateful that we had very efficient ground staff who maintained my plane to a very high standard.

Vic Gill

GROUND CREW

Flight Lieutenant Bill Astell was an exceptionally friendly fellow – but I got to know his flight engineer better because he was interested in the mechanics of the aeroplane. Sergeant John Kinnear used to get up on the trestle with us and poke his nose in the engine. He was an amiable sort of chap. Then there was Squadron Leader 'Dinghy' Young. He was very officious. He always aimed for perfection and often had his aircrew on the site checking instruments. He was a little aloof from us, to some extent. Some of the other pilots were more like one of the lads.

Sergeant Ray Grayston

FLIGHT ENGINEER, AJ-N

It's amazing to me that the Lancaster stayed in one piece, the way we had to fly it. When you look at how modern aircraft structures are made, there's no comparison at all.

It was quite an easy machine to manoeuvre, the wings flapped about a lot and the engine flopped up and down a lot, and when they'd done a lot of hours they were really a bit of a bag to fly, but we were lucky because we always had pretty new machines – they degraded rapidly when they'd done lots of hours.

How important were ground crew? They were super.

Les Knight was an exceptionally good pilot – a brilliant pilot – and he'd fly the Lancaster like a fighter. Old Les would come in on the wing and drop it down on to the runway.

We were intensively flying on ops, no bullshit and no brass-polishing your buttons, no parades. You got up in the morning and flew and you went to bed, and that was it.

The gunners were good at their job, Hobday the navigator was brilliant, Johnny was absolutely brilliant because he was a navigator, and he volunteered to do a bomb-aimer's job, so we were dead lucky in our crew – we had two navigators in the crew and the pilot was a navigator – and that's why I'm here today.

We were there to do a job. Gibson made that very clear that we weren't there to play silly buggers.

Flight Sergeant George Chalmers

WIRELESS OPERATOR, AJ-O

We flew all over Lincolnshire, and there was one particular farm, and every time we took off from Scampton, especially in daylight, round about lunchtime – we used to go down to the 100-foot level. Right on our track was this haystack that these land girls were building and we used to zoom over them. But this particular day we had got a bit off track, we came over the farmhouse itself and a couple of horses leapt over the gate and dashed across the field.

People were complaining all the time to Scampton. Even people *at* Scampton complained, as we were training over the airfield as well.

We trained everywhere – but mostly over canals and reservoirs flying at 60-foot level. The low flying could be just about anywhere, different places, to complete a trip. The idea was mainly to get our bomb-aimer used to the different countryside and the obstacles in the path – electric pylons, any high wires, buildings – so that he would know the terrain in front of him before he went.

One of the canals we trained over led into the Wash, and there were one or two in Derbyshire. The lower-level flying was another aspect of the job – because we had to do this all the way to the dam and back. We had to get used to the idea of flying over land at that height with all these different obstacles in our path.

Flight Sergeant William Townsend

PILOT, AJ-O

Initially the low-level flying involved flying the operational Lancaster as if we were on a flying training unit.

Then we had refresher courses on the link trainer – blind-flying training. The rest of the crew worked in their own particular capacities as gunner, navigator or wireless operator. They had to keep up to date with all modern improvements in equipment and we had to practise all these things on the ground and in the air as a crew.

The link trainer was a static aeroplane. It was a box with a lot of instruments inside which was controlled by an instructor outside. It had a control panel and it was similar to flying with a blacked-out cockpit at night where you just had to learn to fly by the instruments in front of you. You had to steer a course across country, and you'd meet all sorts of different conditions such as bad weather, changing winds and you'd have to learn to fly this thing on the ground until it came naturally to you.

Flight Sergeant Leonard Sumpter

BOMB-AIMER, AJ-L

We all got to know each other – our crew – and we were assigned to our plane. Then we started off on low-level cross-country flights – sometimes one in the morning and one in the afternoon – or sometimes one in the afternoon and one at night. You just progressed from there because nobody knew what was going on. It was just a matter of training all the time.

Flight Lieutenant David Shannon

Pilot, AJ-L

We still had no idea why we were being trained or what the target was going to be, but the three major points made to Gibson were that we had to fly in formation, at low level – and master what literally became map-reading, because the navigational aids which we had at the time were not very effective at low level. We were flying mostly overland apart from crossing the North Sea into enemy territory – and it was pinpoint flying from one strip to another. We had to map out eight or ten different cross-country routes, flying time about three-and-a-half to four hours, and all the navigators made bearing strip maps of the sections, so that we did virtually all of the navigation by map-reading on these flights.

Flight Sergeant Leonard Sumpter

Bomb-aimer, AJ-L

We'd work between turning points on legs of about fifty or sixty miles, so you weren't staying on the same course for hours. You'd do fifteen minutes this way, and fifteen minutes that way – dog-legging to keep you sharp. This was good training.

Then it got harder. We did some night low-levels. If it was pitch dark you just couldn't see to map-read well – so in the daytime they brought out a scheme where the pilot's windscreen and the nose were covered in a blue material – the planes went to Waddington for the screens to be fitted. The pilot and bomb-aimer wore tinted glasses, which made it look like moonlight outside – which would be the conditions for the night of the raid.

This way you had simulated moonlight – which was quite bright – and you could do your map-reading very well from low level as if it were night. When it was pitch dark it wasn't any good trying to map-read from low level, as you just couldn't see anything. You could see rivers and major landmarks, but you'd got to see four or five miles ahead all the time. When you got out of the plane after the exercise, everything looked coloured for a while and you had to put on dark glasses until your eyes became accustomed to normal conditions.

Flying Officer Edward Johnson

BOMB-AIMER, AJ-N

We worked to get used to flying at low level in very big aircraft – in daylight, then at night. At first this was simulated night with screens blocking the natural light and wearing dark glasses, leaving just one fellow who could see out in case anything went seriously wrong. This was usually the flight-engineer. He didn't wear the glasses so he could see out in case anything came up that the pilot hadn't seen, so he could warn him of any danger. Then finally we went flying at night – and then in formation at night. As far as I was concerned, I had to learn a new bombing technique, because there was no bomb-sight of any consequence that could operate at the height we were going to have to operate. It meant doing a lot of low-level bombing to develop techniques to drop the bombs – which was made more difficult because we didn't know the target.

Wing Commander Guy Gibson

We tried to get practice by judging height. To do this we put a couple of men on the side of a hill overlooking a lake with a special instrument to measure aircraft height. One by one we dived over the water and were told afterwards whether we had been right or wrong. This was all right for daylight, but proved impossible at night. Then one day the problem was solved. Mr Lockspeiser of MAP paid the SASO a visit. He said, 'I think I can help you.' His idea was an old one – actually it was used in the First World War.

He suggested that two spotlights should be placed on either wing of the aircraft, pointing towards the water where they would converge at 150 feet. The pilot could see these spots, and when they merged into one he would then know the exact height. It all seemed so simple, and I came back to tell the boys.

A party of men descended on our workshops and all the aircraft were fitted within a matter of days. Night after night, dawn after dawn, the boys flew around the Wash and the nearby lakes and over the aerodrome itself at about 150 feet, while others with theodolites stood on the ground and measured their heights. Within a week everyone had got so good that they could fly to within two feet with amazing consistency – but as I stood there and watched them with their lights on, I knew that whatever losses there might have been before, we were certainly going to make it easier for the German gunners flying around with lights on.

Night flying

Gibson's crews flew round the clock, using all the available hours of darkness as summer nights drew out to hone their night-flying skills. The squadron's aircraft were adapted to simulate night-time conditions in the cockpit – but at such low level, the risks of night flying were enormously increased.

Flight Lieutenant Les Munro

Pilot, AJ-W

In some respects the amount of flying we did over water prompted some speculation – there were not a great number of crew members who, from the training over the lakes that we did, decided that we were going to attack the dams – it was not until the afternoon of the attack itself and the briefing that the majority of crews realised what the target was.

Flight Sergeant George Chalmers

Wireless operator, AJ-O

Night flying was even more difficult for us, but we tended to take the familiar routes – ones we had done before over and over again. But they weren't long sorties, they lasted about two or three hours at the most. It was more acclimatisation than anything and getting used to flying at that height.

Sergeant Dudley Heal

Navigator, AJ-F

We trained for six weeks – low level all the time – and this in itself aroused a great amount of interest and also some worry because flying at low level in a Lancaster with a bombload is an ordeal, I would think, for the pilot. It required very careful observation of the land below and also, because a lot of it was over water – reservoirs, rivers, canals, the Wash – it got a lot more difficult at night. Night flying low level over water – at about 200 feet in something like the Lancaster – well, if the water's smooth you've got a job to see it in the dark. It's very easy to fly into it. If it's rough the sea shows up better.

Steve, the bomb-aimer, and I worked very close together, and we had all these various training operations – which was great practice for us.

Wing Commander Guy Gibson

The synthetic night-flying aircraft had arrived; we all had a crack, and found it perfect. It was funny, though, flying along at night in daytime. It makes you quite sleepy. The boys had begun to get good at their flying, and now I knew where and what the targets were, we could plan a route similar to the one which would actually be used over Germany. This meant flying a lot over lakes, but the excuse for these was always the same: they were good landmarks and a good check on navigation. As we would have to fly to Germany at tree-level height, keeping to track was most important, and it meant navigation to the yard. There hadn't been any accidents – yet – but birds were already beginning to get in our way, coming into windscreens and getting stuck in the radiators; trees were being hit and sometimes even water. Many times out at sea the boys were getting fired on by His Majesty's ships, who have notoriously light fingers, but they took it all in good stead because, as Micky Martin put it, 'It's good practice. It makes us flak happy.'

Security

Any leak, no matter how small or apparently insignificant, could endanger not just the success of the mission, but the lives of the aircrews. Security was taken very seriously and no-one was to talk about the nature or circumstances of their training. Operating on a 'need-to-know' basis, none of the crews knew of their target until the pre-raid briefing.

Kenneth Lucas

GROUND CREW

Nothing was known of the purpose of this squadron and this was on everyone's mind – why are they doing this? There was top security. Some of the ground crew lived out in Lincoln, and apparently they were discussing things on the bus on the way into camp, and they were hauled before the security people because of what they'd said. Our mail was monitored. I don't know about the aircrew, but we had no idea at all.

Flight Sergeant Leonard Sumpter

Bomb-aimer, AJ-L

57 Squadron was next door, but we never let on to them what sort of training we were doing. They probably guessed, though.

There was a joke around Lincoln – there was a hotel called the Saracen's Head, and the joke, which proved to be true once or twice, was that if you went in there for a drink at lunchtime, the barmaid would tell you what target you were on that night. I don't believe it myself, but some say that she probably guessed right once or twice – but that was a standing joke.

Flying Officer Harold Hobday

Navigator, AJ-N

There was very strict security about our training – we weren't allowed to talk about it at all. Although we didn't know exactly what our target was. Everybody in Lincoln must have known we were low flying – much to a lot of people's annoyance on the ground, because it's a bit noisy when you get a few Lancasters going along at 100 feet. But we all obeyed the tight security rules – as far as we aircrew were concerned it was our necks that were involved. And I don't think the ground crew knew very much of what was going on except that it was low level.

Flying Officer Edward Johnson

Bomb-aimer, AJ-N

There were no particular measures taken to maintain security on the squadron, except that we were told not to talk to people – and we didn't even say anything to our families. We were particularly not to talk about the type of training we were doing, and I think everybody was sufficiently experienced to observe that. Most of the crews were experienced at being shot at, and knew what might happen if they didn't keep security.

Flight Sergeant George Chalmers

Wireless operator, AJ-O

Security was very, very tight – it was so tight that we had people listening to us in pubs and places like that in town, in Lincoln, and repeating our conversations to us and telling us telephone calls were being recorded, if you made any. It was quite surprising sometimes what people actually said! For instance, they'd say, 'I won't be coming to see you this weekend because of this that and the other' – from which you can construe a lot of information. It was carelessness – a slip of the tongue mostly, rather than any intentional leak.

They told us that these 'spies' were monitoring us. They didn't tell us who they were – but they told us it was being done. We had the tannoy system which indicated that our squadron had to go to the ops room. We went up there and that was the first time that we discovered what they were doing, and they read out statements to different people who had said this, and we were quite surprised to know that we were being recorded, or even overheard. And some letters were also censored too. It was altogether a stern warning – be careful what you say.

The security people were probably from London, and they just made their reports and passed them to the squadron commander. When he got them Gibson was a bit surprised and called the squadron together to read these statements out, much to the surprise of the individuals who said them! More of a laugh to us, but at the same time it was quite serious.

Wing Commander Guy Gibson

The security measures were evidently efficient. All our telephone wires were tapped, all conversation checked. One boy rang up his girlfriend and told her that he could not come out that night because he was flying on special training. Next day, in front of the whole squadron, I told him that a lapse of this kind would end in a court martial. There was no more loose talk.

Flight Lieutenant David Shannon

PILOT, AJ-L

We were informed that security had to be tight. There was one chap who phoned through to a girlfriend and said he was going off on a low-level cross-country flight and he was hauled up in front of the entire squadron by Gibson and told that, although nothing had come of this, this was the sort of breach of security he didn't want, and that nobody was to talk about what they were doing to anybody. The example he made of that poor bloke stressed the importance of security, and we heard no more of any breaches of security from anybody. It was as tight as a drum. I think Group had also posted in quite a few additional security people around the aerodrome, ostensibly doing some other jobs, about whom we knew nothing – but they were there to ensure that there was no breach of security.

Flight Lieutenant Harry Humphries

ADJUTANT, 617 SQUADRON

The security officer at Scampton, Flight Lieutenant Evans, I believe, did a grand job of work. His job was made easier due to the magnificent co-operation of both the ground and aircrew of the squadron. Through six weeks of rigid censorship there was not one instance of a breach of security at Scampton.

Bomb-aiming

One basic bomb-aiming device was developed for the mystery raid, and the crews adapted it to suit their way of working. Their brief was to train to achieve pinpoint accuracy at the designated height and speed. The crews trained relentlessly with standard practice bombs to hone their skills.

Wing Commander Guy Gibson

One day in April a wing commander from MAP came up to see me. I was sitting alone in my office when he came in. He began at once to tell me of our sighting difficulties on the Eder and Möhne dams. The latest anti-submarine bomb-sight, he explained, would be no use when carrying out our special form of attack. I was horrified, but he went on to explain. He was the sighting expert and had been let into the secret so that he could help us. No-one else knew.

Between us we worked out a plan, although it was his original idea. He took out a piece of paper and by drawing queer lines he explained what he meant – it took the form of a very simple bomb-sight using the age-old range-finding principles. From aerial photographs we had noticed that there were two towers on the dam, and when we measured these, we found they were 600 feet apart. Our mines had to be dropped exactly a certain number of yards short of a certain position along the dam wall. He worked it all out. He was a mathematician. Then we handed it to a corporal with glasses, and within half an hour the instrument section had knocked up the prototype of the bomb-sight. It cost a little less than the price of a postage stamp.

Soon, using these sights, we were achieving extraordinary accuracy – but then we went on to night training. Again I found it very much harder, because although we could see the targets clearly enough, we couldn't see the water. It was practically impossible to fly at exactly 150 feet. A few went too low and rubbed their bellies along the water; others went too high and under-shot by miles. After the second night's bombing, pilot Squadron Leader 'Dinghy' Young landed, sweating. 'It's no use. I can't see how we're going to do it.'

Flight Sergeant Leonard Sumpter

BOMB-AIMER, AJ-L

We used to go dropping practice bombs from low level at Wainfleet from about 60 feet with no sight, just to keep your eye in. Some of the cross-countries were pretty long – round the north of Scotland, down the Caledonian Canal and off the northern coast of Ireland. The main exercise in low-level flying was to sharpen up our map-reading, because at high level, navigators can give an easy change of course because you can see an aiming point six miles before you get to it. But at low level, if you don't hit your aiming points, by the time the navigator gives the pilot a correction you're about ten miles past where you should be turning. So it was up to the bomb-aimer to keep the plane on course as much as he could – which he did by not telling the navigator, 'You want a correction here.' Instead you'd see how much you were off course and you'd tell the pilot, 'Five degrees to port' or 'Five degrees to starboard', until you were back on track again, then you'd tell him to resume his course. Low-level flying was mostly up to the bomb-aimer and his map-reading.

Sergeant Dudley Heal

NAVIGATOR, AJ-F

The next unusual thing was that we did a lot of our training over reservoirs, particularly Derwent Water. As time went on, we practised bombing where you pretended to drop a bomb at a certain distance before your target and at a certain height. We had no idea why – normally you'd fly over something and drop your bomb on it. Then they brought in a new instrument which Steve Oancia, the bomb-aimer, held and, looking back, it was configured like the dam itself. It was a piece of wood with two arms and as he held it, when these two bits of wood lined up on two markers – the turrets on the dam – that in effect was where we would pretend to drop our bomb.

Wing Commander Guy Gibson

After a few days' low-level training I got the pilots together: 'I've been speaking to the maintenance people, and they tell me that already a few aircraft have come back with leaves and tree-branches stuck in their radiators. This means the boys are flying too low. You've got to stop this, or else someone will kill himself, and I might also tell you that the Provost Marshals have already been up to see me about reported dangerous low flying. We all know we've got to fly low and we've got to get some practice in, but for God's sake tell your boys to try and avoid going over towns and aerodromes, and not to beat up policemen or lovers in a field – because they'll get a rocket if they do. Now if you'll excuse me, I've got to do some trials on a certain reservoir, to find out if we can do the thing at all.'

Half an hour later, we were on our way in my faithful G for George towards the reservoir called Derwent Water. This lake is in the Pennines, surrounded by high ground with just enough industrial haze blowing over it to make it ideal for the job; moreover, the water was always calm because there was no wind in that valley. Remembering what Jeff had said, we came screaming down at the right air-speed to pull out as near 150 feet as we could judge – Hoppy was sitting beside me – and released a missile. It fell short. Then we were twisting and turning our way through the valley, with high hills on either side. We climbed up again for several more tries, and in the end found it more or less fairly easy.

That night at dusk, with the fog already beginning to fill up in the valley, cutting visibility down to about a mile or so, we tried again. This time it wasn't so good. The water, which had been blue by day, was now black – we nearly hit that black water. Even Spam said, 'Christ! This is bloody dangerous,' which meant it was. Not only that. I said to 'Dinghy' there and then, that unless we could find some way of judging our height above water, this type of attack would be completely impossible.

Flight Sergeant Ken Brown

PILOT, AJ-F

We started out using the Derwent Dam among other targets – and believe me we didn't have a clue as to what was going to be the target. Nobody even mentioned dams – we thought it was going to be the battleship *Tirpitz* or some other thing.

We went up to the Derwent Dam and there was moonlight, but unfortunately there were a few clouds around. And in the Derwent there's a row of hills down the east side and a slight cut-off at the end. You've got to cut around this to come at the dam.

Guy Gibson decided he would make the first run, and his bomb-aimer happened to be an Australian. So he runs in on this dam and just as he was going in a cloud came across the moon. So it was damn dark. He came down on the water, without lights, then went rushing towards this thing. We were equipped with VHF radio, which was the fighter boys' radio and there was a toggle switch at the side. Transmit was one way; receive the other way. Guy left it open, on transmit. So as he dashed in towards this dam, the bomb-aimer says, 'This is bloody dangerous!' I think everybody in every aircraft was harbouring those same thoughts.

F/S Ken Brown

F/O Edward Johnson

Courtesy of www.ww2images.com.

Sgt Basil Feneron

Sgt Dudley Heal

Flying Officer Edward Johnson

BOMB-AIMER, AJ-N

We worked to get accuracy, dropping the bombs at low level by eye, judging the distance. Our height and speed were pretty well fixed, so you got used to gauging distance for releasing the bomb. This changed as soon as we knew the type of target – we were able to think about how we were going to make these distance judgements.

The way they worked out was to have a T-shaped piece of wood with two pins on the end of the T and an eyepiece on the stalk of the T which formed a triangle which represented the triangle made by the towers on the dam walls. I personally didn't use this method – I found it clumsy and inconvenient and not very accurate. I developed my own technique of using a long piece of string, fastened each side of the clear bombing panel, and some marks on the panel itself in grease pencil, which did two things when the time came to do the raid. It enabled me to have two marks – one to suit the Möhne tower distances and another to suit the Eder tower distances – which weren't the same.

I believe it goes the rounds that Flight Sergeant Sumpter also claimed to have invented the string device, but I think he was a little bit later than me using it. It worked very well because I had a bigger triangle than the official one, and it was stationary rather than having to be held.

Flight Sergeant Leonard Sumpter

BOMB-AIMER, AJ-L

Our bomb-aiming equipment was a piece of string – with a knot in the middle. There was a chap named Wing Commander Dann, and he invented a triangular piece of wood with a handle underneath, an eyepiece at the back and two pieces of wood on the front. When you held the eyepiece to your eye, you looked through it and saw the two nails at either end of the piece of wood. When these nails coincided with the towers, that was the time to release your bomb. But we found, when we were trying it out, with the movement of the plane and your finger on the button, you had nowhere to lean – you couldn't steady yourself. So I and a fellow named Johnson, who was in Les Knight's crew, we talked it over and we came up with a solution that if you had a piece of string round the clear vision panel. On the Lancaster were screws which held the Perspex in – and if you had a piece of string fixed to two of the screws opposite on the Perspex panel and brought it back to your eye to make the triangle – which was the same as the wooden sight – you could work your distance out from the screen to your eye in proportion to the width between the towers. This worked just as well. And you could lean your arm on the arm rest, and just lie there with this against your eye, without wobbling. You had your other hand on the button, whereas before, holding the piece of wood up to your eye, you were very shaky.

A lot of people used the string and the chinagraph pencil. On the clear vision panel we had two marks, about eight or nine inches apart, so that when you held the string back to your eye, the distances were in proportion. The width between the towers and the distance before the dam at which the bomb had to hit the water were worked out on that triangle. Each crew had a choice as to what equipment they'd use – just what worked best for them.

Flying Officer Harold Hobday

NAVIGATOR, AJ-N

We did a lot of practice with 11½ lb practice bombs. Then they rigged a couple of markers on a lake, because eventually the range for dropping the bomb was calculated by triangulating a device in the aircraft with the two towers on the edges of the dam to give the right bombing distance. These points were rigged up – flags –

so we could practise the triangulating process. This was done, in our case, by a piece of string which the bomb-aimer held to his eye, and two marks on the Perspex in front, and when the Perspex marks in front coincided with the two flags, that was the distance to drop your bomb. Gibson certainly seemed to be satisfied with our progress with target bombing – we did quite a lot of practice, and God knows how much fuel we used on this, but it was well worth it.

Flying Officer Edward Johnson

BOMB-AIMER, AJ-N

The bomb practice didn't involve a lot of bombs. We used to drop a few practice bombs at Wainfleet Sands in the Wash, but it was mainly concerned with practices to position the aircraft in the right place to release the bomb than the actual dropping. This was the major preoccupation, both with the flier and the bomb-aimer, who were trying to make the bombing possible. The pilot had to do his bit, the engineer had to keep the speed constant, and I'd got to assess the time to release the bomb.

Flight Sergeant Leonard Sumpter

BOMB-AIMER, AJ-L

Between the cross-country flights, to give us a bit of a diversion or entertainment, we'd take half a dozen practice bombs and go out to Wainfleet, where they'd stuck a post in the marshes. We'd see if we could hit this post from 60 feet – anything to break the routine, because the flights did become a bit monotonous after a time.

When you thought to yourself that you'd got to a peak and that you couldn't get any better as a crew, that was when you most had to keep up your concentration. Every time you had to do the flight as though it was your first or second, however experienced you were and however well trained. Otherwise, if the pilot didn't concentrate, you'd probably hit a tree or a pylon. We were only at 60 to 80 feet all the time on the cross-countries – which were routed to miss mountains and high ground. All the same, one or two chaps came back with branches in their air-intakes.

From experience on these flights we decided that the bomb-aimer should give the directional corrections and pick up the track – so that if the heading wasn't quite right and you were only a little bit off track, it was better to keep that course from one point to another and give slight corrections. Then we'd go back on the same course again, rather than change the course which might throw you off on the other side of the track.

Down low, the winds tend to vary from area to area and so this was the best way to do it – to give the corrections so that if you were, say doing a sixty-mile leg, if the navigator's course wasn't quite right, you'd probably do two or three dog-legs in the course. You'd go straight and then you'd veer to port and straight again – then veer to starboard and straight again. As long as you kept on track, that was all we worried about.

Wing Commander Guy Gibson

By the end of the third week, all crews had done about twenty cross-countries by night, and could now find the 'tree', which I had asked them to find. Their navigation was really expert. We had dropped about 1,500 practice bombs on our range, and the average error for all these was as low as 25 yards. With this basic training behind us, we felt that the time had now come to plan our route to the target, so that we could imitate it as much as possible over this country.

Super-low flying

The crews were mastering the art of flying, navigating and hitting a target from 150 feet – but then the goal-posts moved. The word from the boffins was that the new bomb could not be delivered from 150 feet: the flying height would need to be just 60 feet.

Wing Commander Guy Gibson

When I got back to Scampton from seeing 'Jeff' at Brooklands, we immediately altered the two spotlights so that they converged at 60 feet and made our first trials over the Wash. I think Maudslay was the first to take up the aircraft to experiment at this height. He found it all right, but said that it seemed very low. We heard that these spotlights would work satisfactorily over water with a slight chop, but over glass-calm the spots would shine through the water and converge underneath, so that there was a very great danger of flying straight into the drink without knowing the danger.

Flight Lieutenant David Shannon

PILOT, AJ-L

Unbeknown to us at the time, Gibson was having meetings with Group and the specialists – particularly the research departments – and he was let into the secret, I suppose, towards the end of April, as to what the target was to be and what bombs we were to use. Experiments were being carried out, dropping the mines to see how they performed. They found that dropping them from a speed of something like 240 mph from 150 feet, the casing just shattered. So Barnes Wallis had to keep working on this. They came up with the solution that the only way, if they were to carry out the operation, was to fly at 232 mph and drop the mine from 60 feet.

Flying at 60 feet in daylight is a very different kettle of fish from flying at 60 feet at night – and the only way we could fly at that sort of level without killing ourselves was with a very accurate altimeter. There was no altimeter sufficiently accurate available at that stage to allow us to contour fly at 60 feet, so the research people put a spotlight in the nose of the aircraft and one in the tail, and set the angle of the beams so that they converged to one spot at an exact height of 60 feet below the aircraft.

The navigator looked out and guided the pilot. We practised doing that quite a lot at dusk and dawn over the aerodrome runway. It was fairly low for night flying; you'd be going across the middle of the aerodrome and the navigator would be saying 'down, down, down' and you thought he'd never stop. You just had to accept that he knew what he was doing. We were more accurate over the water. We were flying over the Wash, getting down as low as we could, and were aiming and hitting targets 6 feet in diameter with practice bombs.

We found that actually flying over the water we could be very much more accurate over still water than we were over the runway. It was very successful – but it was hellish low – especially in the first runs we made. It was essential that the bombs had to be released at 232 mph – so we made a little red mark against the 232 on the air-speed indicator, so that when the needle came round, the pilot could see it. Then the navigator would move out of his seat and look through the blister to control the height.

All this was not so much dangerous as likely to cause damage to the aircraft. There was a certain amount of trouble with branches from trees, and some damage – but nothing serious. Birds were a problem too – we had quite a few aircraft damaged through hitting birds at that height. It's surprising the damage a bird can do to an aeroplane. We were propeller aircraft, so there was no real danger of damaging the engines – they were just smashed up. But a bird would come through the fuselage or through the wings, just like a bullet, making a hole.

We practised formation flying to give us a pattern against any attack that might come from the fighters. The bombers of the Second World War were only equipped with .303 machine-guns, and to concentrate fire-power it was more sensible to have aircraft in vics of three to give us stronger fire-power than individual aircraft, and to try to keep the aircraft together, because the whole secret of the operation was to be surprise, and at this low level, the accuracy.

Flight Lieutenant Les Munro

PILOT AJ-W

The experience of low flying was not something that was usually carried out by bomber crews, and it was normally taboo and a court-martial offence if you were caught doing it unless you were authorised. But the crews enjoyed low flying – considering it was authorised and we were programmed to do it. Piloting at low level was largely a question of judgement and of being able to determine how close you were if you were at really low level and you had objects approaching you like tall trees. You had to become adept at gaining height before you clipped the tops of whatever you were approaching. That was the major part of low flying as far as the pilot was concerned – being able to judge distances at 200 odd miles per hour.

The other major aspect of our training was navigation, which involved all crew members. The navigator was primarily responsible, but it involved becoming proficient at identifying landmarks such as small towns, crossroads, particular buildings and for the other crew members to become proficient in map-reading, so they could tell the navigator what was coming up; he was the focal point, checking the route on the map. When you're flying at 60 feet – sometimes lower – you were approaching landmarks at a terrific rate and you had to be able to identify them quickly as opposed to flying at say 15,000 feet, when visual navigation was relatively simple – you could see for so many miles ahead and you had time to distinguish any particular landmark. That was the difference.

Sergeant Ray Grayston

FLIGHT ENGINEER, AJ-N

I enjoyed it personally – it was exciting flying at ground level in a Lancaster, it's like riding a motorbike at 100 mph in those days. I was responsible to back up the pilot on controls if he needed it – throttles, flaps, because he's sitting on the port side of the cockpit and he can't get to the starboard side. All the emergency buttons were on my side – engine feathering, fuel control, and there was an engineer's panel down to the right with about forty instruments, and you had to get up off your arse and sit down and log, every twenty minutes, temperature, fuel, in other words everything in prime condition, the engine not losing oil or overheating. You sat up by the pilot on the bench seat. The pilot's got a nice cushy seat, but the engineer's got a seat you could hardly sit on. You pull out a footrest for your feet – extend the bar – the engineer's seat is a drop-down affair with a strap back and you just sit on it. But because of the job you were mostly on your feet anyhow – if the rear gunner was in trouble you'd go down and see if he was fit, and you'd get in contact with the crew if the intercom failed, you were on the move most of the time so I didn't spend much time on the seat – in fact I spent most of the time sitting on the bloody floor.

Derek Dobbs

AIRCRAFT RIGGER

I often went on practice trips – they were very exhilarating. The only space available was usually standing behind the pilot, and flying low level in a Lanc is an experience – especially over water. Quite a lot of bird strikes occurred because of the amount of low flying.

Bill Townsend was a good pilot. One flew with him with great confidence. I think it gave them confidence in our maintenance of the aircraft if we flew with them.

Flying Officer Edward Johnson

BOMB-AIMER, AJ-N

We had a few scrapes in training – like picking up treetops. That was fairly common when flying so low. But this didn't do much damage to the aircraft – sometimes a few branches got sucked up oil coolers, but it was nothing serious. The low-level flying was frightening at the beginning, but it's surprising how accustomed you become to these things. You begin to think it's normal and you adjust yourself to the new circumstances. We didn't consider it hazardous by the end.

There were certain areas that were stipulated by the Air Ministry as being OK for low flying, so we did quite a lot of preliminary bombing training at the Uppingham Reservoir in the Midlands. Sometimes there were complaints from the public – but never to individuals – rather to the squadron.

There weren't generally any flying pranks – it was a fairly free and easy squadron, but we all knew that we were doing a job and while we were doing it, we were serious. Afterwards we were a light-hearted squadron – lots of fun and talking and maybe a few pranks in the mess.

The training was serious – and interesting to the nth degree, because it was progressive. You started high and got lower, then you start formation flying – which in itself is quite hair-raising at the beginning. It was fascinating, right to the end.

Flying Officer Harold Hobday

NAVIGATOR, AJ-N

It was down to the pilot to use his eyesight to ensure that we didn't hit the ground – before this operation we didn't have any special equipment to assist low-level flying. Eventually we had lamps which were shone down on to the water which converged and told you when you were at the right height – but normally we didn't have anything like that. I think the pilots found it all right over land, because you see things which help you estimate your height – but over the sea it was very difficult – and one of our crews actually hit the water going over the Zuider Zee because you couldn't tell the height over water very well. But the system of lamps worked marvellously – it was first class. It was my job to look at the spots – I looked through a bubble in the side of the aircraft on to the water and when the beams coincided, we were at the right height. I had to say 'up' or 'down' to the pilot – it was a marvellous idea. We practised with it once or twice before we went on the dam raid, but before that we didn't have it at all.

A lot of people didn't want us to use the system because the German gunners would be able to see the lamps and could take a better aim on us – but we had to risk that because it was essential to have the aircraft at the exact height of 60 feet when we dropped the bomb. As a further adjustment, all the guns were altered. Normally they had a tracer every two or three bullets – but ours were all tracer, to make it look more fearsome to the Germans. And I think that worked.

We trained to switch the two beams on as soon as we got down over the water – just before we started the bombing attack. We didn't go round in the sky with the lights on.

Sergeant George 'Johnny' Johnson

BOMB-AIMER, AJ-T

McCarthy was six foot plus, and almost as broad as he was tall, hands like hams, and they really held that aeroplane wherever he wanted it to go. At Sutton Bridge there are some electric cables which went across the canal then over the bridge, and we used to fly underneath them. We weren't supposed to of course, but it was an added buzz and it was a bit of daredevil stuff, and we were young and stupid at the time, and nobody got hurt doing it – so we just went on doing it. On one occasion we ended up at the bombing range to do a practice run, and we were flying at 30 feet, and somebody flew underneath us.

Flight Lieutenant David Shannon

PILOT, AJ-L

The flights took us all over the country – up as far as Scotland and all through the Midlands, with quite a lot of practising up round the hills around Sheffield and down as far as Wales – but once the various cross-country routes had been planned, all the stations and patrols and anti-aircraft systems in the country at the time were warned about it, because in the early days there were nothing but complaints and confusion coming in to Bomber Command Headquarters – everybody was sighting Lancasters flashing all around the country at very low level – and what the hell was going on?

It was great sport on a summer day to fly as low as one could down the canals over Lincolnshire and upset the boats and people punting – and see them scuttle for the shore. With hindsight, it was probably not as much amusement to them as it was to us at the time. However, it was a very serious time, and we were training seriously for we knew not what – other than we had been told that if it was successful, it could well lead towards the shortening of the war, so everybody during wartime was fairly serious about these things. The flying and training were a very serious business.

Sergeant Basil Feneron

FLIGHT ENGINEER, AJ-F

We were always a very good crew. Ken Brown always kept us up to scratch. He used to take us away to knock seconds off dinghy and escape drills, while a lot of crews were more interested in the birds they were taking out. It was the continual training that helped us to survive.

With 617 Squadron we had weeks flying around the country, flying low, hugging the ground, sometimes coming back with leaves and twigs wrapped round the rear wheel, which was not retractable. We got the odd complaint. A farmer or two were blown off the seats of their tractors – not from the draught of the wings, but sheer terror.

We went over the Trent, nipping over barges, and flew under high-tension wires with Ken calling, 'Come on, there's plenty of room!'

Flight Sergeant George Chalmers

WIRELESS OPERATOR, AJ-O

We all took part in all the training, every time the aircraft flew out. We did training out to the North Sea and back – all at low level. We tried to keep it at certain periods of the trip, at just 60 feet above the water.

This low flying was more exhilarating than nerve-wracking. I quite enjoyed it. To see things rushing up towards you all the time was quite an experience. I suppose other people might have been a bit upset by it but I was sitting in the aircraft, and it didn't seem too dangerous. The bloke right in the nose perhaps, the observer, bomb-aimer – he would probably get a bit of a fright now and again, especially when he saw a pylon or electric wires come zooming up towards us.

We trained all over the country – we only went to the North Sea to get away from the local people who were objecting to us flying over them all the time, with these aircraft bellowing away at four engines, day and night. They were a bit frightened of us, I think.

Flight Lieutenant Les Munro

Pilot, AJ-W

I'd say the low-level flying was exciting – exhilarating. It wasn't terrifying unless you didn't have any confidence in your ability to handle the speed at which you're approaching structures.

Sergeant Basil Feneron

Flight engineer, AJ-F

Our main concern was high-tension cables, it was pretty light, but those were difficult to see at times, and we'd got our mikes on continuously, all the time, and you'd just say 'High-tension cable coming up,' and he'd say 'OK, I can see it' – and we'd nip up, get over it, and get down again.

Flight Sergeant Ken Brown

Pilot, AJ-F

The high-tension cables would twinkle in the moonlight, and if you could see the high-tension wires there – you know you can go over them – but if you see them just that much higher – you can't. You've got to make a decision and go under them. You'll never make it over. You've got to make that decision long before you reach that. People saw them up here, and tried to go over them, and just smashed through them. Bad news.

John Elliott

NCO ground crew

With regard to Harold 'Micky' Martin, some people may have said he was a daredevil, but I would say he was an extremely efficient low-flying pilot. One day, from the control tower, I saw him fly in a Lancaster between the control tower and the hangar and we were looking down from the control tower on to the top of his Lancaster as it went past – and I'd say that's some low flying. He couldn't have been more than 20 feet from the ground.

Martin had the reputation of being very good at low flying. But I think all the pilots in the squadron were good. They had to be the best, and I think they were. We had the cream of the pilots, or crews in fact from the group.

Wing Commander Guy Gibson

By 1 May I had rung up Jeff and told him that we could do it. He asked me to come down for the trials again. Then, one morning in early May, Mutt flew over and dropped one which worked. The man on the ground danced and waved his hands in the air and took out his handkerchief and waved it madly. I threw my hat in the air. I could see Mutt in his cockpit grinning as he banked around after his run, and I waved back at him and shouted into the noise of the engines. I believe, although I do not remember very well, that Jeff threw his hat into the air too. This was a wonderful moment.

Beck Parsons

GROUND CREW

The relationship between the aircrew and ground crew was good – mostly. There was no sense of envy – a lot of the ground crews around were allocated to one aircraft, so they would know that crew very well and they'd be friends, but if you were like I was and covering a number of aircraft, you hadn't got the time to concentrate all your feeling on one crew. One crew flew with Micky Martin, who I thought was fabulous – and he taught his crew all about low flying. I did a trial flight with him, because Gibson wanted that – he wanted all the people to be involved. I flew over the Ladybower dam, one of the practice trips, and Micky Martin, who was the low-flying king, he just got lower and lower and all of a sudden, the rear gunner shouted out, 'Lift it up, Micky, I'm getting soaked.' The tail wheel had hit the water, soaking the rear gunner.

Communications

The specialist nature of the raid and the need for clear communications and split-second reactions rendered the regulation Bomber Command radio telephones inadequate. Gibson pulled strings to have 617's Lancasters fitted with the very latest VHF equipment, previously issued only to Fighter Command.

Wing Commander Guy Gibson

One problem we found was inter-communication. On attacks of this sort there must be no allowance for anything to go wrong, and things had gone wrong here. The radio-telephone sets which we were using were just not good enough. We would have to use fighter sets. I told the AOC that unless we were equipped with VHF, the whole mission would be a failure. I told him that I had been asking for it for some time. He said, 'I'll fix it.' He was as good as his word. Within a few hours a party of men landed on the aerodrome and went to work. Next day the whole squadron was equipped with the very best and most efficient radio-telephonic sets in the whole of the Royal Air Force.

Flight Lieutenant David Shannon

Pilot, AJ-L

A secondary problem cropped up – communications – because we must have communications between ourselves at that time. In Bomber Command the only radio equipment was normal radio control on a medium wave which was used purely and simply for landings and take-offs from base, and then there was radio silence the whole time. The aircraft were not fitted with the right sort of communications equipment, and eventually they got hold of VHF equipment and a special wavelength which we were able to use, similar to the type of equipment fitted in fighter aircraft. This worked quite well, but we had to have quite a bit of practice on it. Our main problem was not so much with using it, but with resolving a code to restrict the amount of traffic over the VHF R/T to stop people from conducting any sort of conventional talk over it. We had to devise our own codes for the operation.

Sergeant Ray Grayston

FLIGHT ENGINEER, AJ-N

They replaced our radios with the current fighter equipment, so that we could communicate up to 120 miles, that was essential because Gibson had to be able to talk to us, and give us instructions as he was in charge of the raid, and we had procedures we had to comply with. There was also that we had to communicate with Gibson and he with us – it was the same VHF radio that the fighters had. And we had a duplicate set so that if the pilot got into trouble, I could communicate as well. It was only used in conjunction with the operation – we were not allowed to broadcast or talk in any other way – it would give positions away.

Flight Sergeant Leonard Sumpter

BOMB-AIMER, AJ-L

We'd communicate through a microphone on your helmet and a microphone strapped to your face, and you had a lever at the side which you put down for speaking and off again when you didn't want to speak. You left it on most of the time because you'd be chatting. You'd say, 'Thrapston coming up about two miles to port, skipper. Go on five degrees to port. Go over Thrapston.' Then Danny would give a new course for the next turning point – that would be Danny Walker who was our navigator – a Canadian.

Flight Sergeant George Chalmers

WIRELESS OPERATOR, AJ-O

These Lancasters were kitted out with very good wireless equipment with a long range – I've used the equipment in the UK and called up Malta in my time. So the range was pretty good.

The transmitter was mounted on a little table in the wireless bay, which is between the two engines – there were two other engines – and behind me were the main spars and above me was the astrodome.

Generally you didn't have any trouble with them. I never had any trouble. If they went wrong, there was not a lot you could do. But you could do things to get a short distance transmission if you had to. You could rearrange the circuit a little. It depended what went wrong you see. The only thing that was going to go wrong, other than a valve going – for which you did have a couple of spares – was a bullet or something that went through the set. But it was possible to rearrange the circuit a little just to get transmission working.

I sat facing the front of the aeroplane. There was the main spar which you climbed over to get to the cockpit, and on the left of that was where I sat, and in front of me was the radio set and further on was the navigator then the pilot and engineer.

I wouldn't say it was cramped, but there wasn't a lot of room to manoeuvre, especially when you had got your flying gear on. We did have an Elsan toilet in the back end of the plane. You had to nip over the main spar and go down into the tail at the end of the aircraft.

The temperature in the aircraft varied according to height and how good the engines were pushing out the heat – a lot depended on draughts. But I wouldn't say it was an uncomfortable ride. I hardly ever sat down anyway. I mostly stood up looking through the astrodome. I only sat down to use the radio gear when I had to.

Sergeant Stefan Oancia

BOMB-AIMER, AJ-F

The navigator was sitting at the table – he could not see outside. Who could see outside were the bomb-aimer, the pilot, and the flight engineer – but they had their work to do, so it was down to the bomb-aimer to do all the map-reading. I would pass that information on to Dudley Heal, and he would plot it on his map.

Sergeant Dudley Heal

NAVIGATOR, AJ-F

Steve Oancia, the bomb-aimer, and I worked very closely together, and I depended very greatly on the information he gave me about what he could see, and that's how we managed to keep largely on track.

Flight Sergeant George Chalmers

WIRELESS OPERATOR, AJ-O

My training was mostly observation, because of the close proximity of aircraft in formation and the low-flying aspect of the mission. Other than normal radio procedures – which I was trained for – I had to get used to other duties which had to be carried out, such as listening in to headquarters stations and sending and receiving; from the communications point of view, we just had certain codes which we used to identify aspects of the raid en route. These were to indicate, for instance, if you were calling off the flight for whatever reason – or they were trying to pass information to you.

We had a list of the code words, the frequencies, and all the call signs. That aspect of the communication was all on a piece of paper. But we were pretty well acquainted with it anyway.

Flight Sergeant Leonard Sumpter

BOMB-AIMER, AJ-L

You could always tell when you were in your own plane – perhaps it was the sound of it or the smell – and in the same way we had our own ground crew. We'd have to borrow planes for training at first, then we got our special aircraft. They'd say, this plane's for Flight Lieutenant Shannon – ED929/G – all those special aircraft had a number with a 'G' – which meant that they should have a guard on them at all times because they were on the secret list.

We did personalise our aircraft – we had Bacchus painted on it because we had so many drinkers in the crew – mind you, they only drank when there was no flying on. If it was a flying day, it was a sober day – and sometimes they didn't have time to drink at night because we were out flying again. But if we had a few days off, we'd go to town.

The modified Lancaster

Each aircraft to take part in the special mission had to be adapted to carry and deliver the new and hitherto unseen weapon. Understandably, the delivery of the modified Lancasters increased speculation as to the nature of the target.

Doug Godfrey

FITTER, AVRO

We knew the Lancs we started working on were for a special raid, but didn't know what. We worked twelve-hour shifts, seven days a week to get them ready. The atmosphere was exciting – electric – and the camaraderie terrific. The rhythm of the workforce was out of this world. Time was so short.

Lionel Dimery

PILOT WITH AIR TRANSPORT AUXILLARY

When we picked up aircraft, as far as we were concerned they were just aeroplanes for the war effort, but occasionally some of them were obviously rather special. I remember one day going to Woodford with a pilot to pick up a Lancaster to take to Scampton. When we got there, this particular Lancaster that I was to assist in delivering had a huge gap underneath the fuselage. The bomb doors had been removed. It just looked an odd shape. I said to one of the ground crew at the time, 'They don't fly like that, do they?' He said, 'Our pilots fly them,' so I didn't argue and that was that. I took three of them, actually, over the next day or two.

I know when we took them over to Scampton they didn't waste any time about taking them off us. We didn't even stop the engines before some of the RAF lads were in them and took them away.

Harold Roddis

FLIGHT MECHANIC

When the new Lancasters arrived we stood out on dispersal and said, 'What the hell is that?' The bomb bay doors and mid-upper turret had been removed. Someone said, 'It looks as if the Lancs

An Upkeep weapon fitted in the bomb-bay of Gibson's Lancaster. Pilots dubbed this ungainly adaptation the 'Pregnant Duck'.

have had abortions.' From that moment, the Dambuster Lancasters were known as the 'abortions'.

Flight Lieutenant David Shannon

Pilot, AJ-L

The practice period went on for about six weeks, during which time the squadron carried out something like 2,000 hours' flying – which was approximately 100 hours per crew for practice training from the formation of the squadron to the take-off for the raid.

It wasn't until early in May that we started getting the modified Lancasters. These had the bomb-bay doors removed and the cut-out section fed off underneath the belly of the aircraft with the suspension arms down to carry these very cumbersome mines, which were the bombs we would eventually drop.

Sergeant Dudley Heal

Navigator, AJ-F

Our aircraft were taken away and when they were brought back the bomb doors had been removed and a framework had been fitted underneath to carry the bomb – but still we had no idea that it was going to be any different from any other bomb.

Sergeant George 'Johnny' Johnson

Bomb-aimer, AJ-T

When the new special aircraft arrived, the first reaction was 'God, do these things fly?' Well obviously they did or they wouldn't have been there. But they had lost the mid-upper turret and the bomb-bay, and in place of that they had this cutaway underneath the belly with a couple of arms sticking down, one either side. We had the same feeling when the bomb arrived. We had no idea what this was all about, bearing in mind that we had no idea what the target was either, or what the operation was going to be. However, we flew these things – and they did fly.

Flight Lieutenant Les Munro

Pilot, AJ-W

The bomb was slung across the aircraft as compared with the normal practice of having the bombs slung lengthways in the bomb bay. The bomb doors were taken right off the aircraft and they had caliper arms on either side of the aircraft which were clamped over either end of the Upkeep – the codename for Barnes Wallis's revolutionary new weapon – to hold it in place. They had a drum on each arm of the caliper arms and these fitted into a flange at each end of the Upkeep. There was a small engine in the front of the bomb bay, which was attached to one of these drums by a belt, and the idea was that fifteen minutes prior to it being dropped, the Upkeep was revolved at 500 revs a minute in an anticlockwise direction.

Flying Officer Edward Johnson

Bomb-aimer, AJ-N

The aircraft were modified so that the bomb bay had been removed entirely and it had been fitted with two hydraulically operated V-shaped arms underneath. These had discs on the end of the arms at the point of the V, which recessed into the bomb itself at both ends to grip it. The action of pressing the bomb-release opened the arms and let the bomb fall out. It also had a pulley on it that could be driven by the hydraulic motor to start it revolving, which had to be done in the air before the attack commenced. This led to some difficulties in flying, because it was like a gyro in effect, and trying to stabilise the aircraft made it very difficult to manoeuvre.

Wing Commander Guy Gibson

One by one the boys put forward their suggestions, and most of them were adopted because they were good ideas from chaps who knew their job.

'Dinghy' said, 'I believe you want the front gunner to remain in his turret the whole time, so that he can deal with any flak guns we may meet. The snag about this is that his feet dangle in front of the bomb-aimer's face. It would be a good idea to fix up some stirrups for his legs, because he would be more comfortable himself and the wretched bomb-aimer wouldn't have to put up with the smell of his feet.' That was a good idea, and the long-suffering Capel was asked to fix up these things straight away.

'Another point,' said someone. 'How about fixing a second altimeter in front of the pilot's face attached to the windscreen so that he can see it easily? It will save him looking down into the cockpit when he's near the drink.' Another good idea, and the instrument boys again were asked to fit up all aircraft by the afternoon.

'How about radio telephony?' asked Dave Shannon. 'I am pretty hazy about this, not ever having been a fighter boy.'

'Well, as far as I can see, this is the procedure,' I replied. I took from my drawer something I had been working on the night before. 'The idea is that we use plain language, backed up by simple code words. If the radio should fail we will have to use wireless transmission (WT) as well. We won't use Air Force code because wars have been lost in the past by the waste of time when decoding. We will invent our own simple code; it must be very simple and self-explanatory. Here's the slip of paper giving the full gen.' On this paper were a few short code words which we were to use. For instance, the word 'Dinghy' would mean that the second objective had been destroyed. 'Nigger' would mean that the first objective had been destroyed. Words like 'Artichoke' and 'Beer' indicated that we would change frequency from Button

A or B. There were many other words, and I told the boys to learn them off by heart.

Upkeep – the bouncing bomb

If the crews were taken aback by the modifications to their Lancasters, they were astounded when on 11 May they saw the weapon they were to deliver. The weight and rotation of the bomb would make their already risky modus operandi even more hazardous.

Wing Commander Guy Gibson

In the early morning, long eight-wheeled trucks arrived on our station covered in tarpaulins. These were our mines arriving. They were glistening new and still warm because they had just been filled with their very special explosive. Our armament officer, Doc Watson, had to deal with these, and it meant a lot of trouble because each one had to be treated like a diamond – they were very delicate and very hard to fuse and prepare. He and his armourers worked day and night getting them ready for our next trials.

Flying Officer Edward Johnson

Bomb-aimer, AJ-N

We were absolutely staggered when we first saw the bombs – it just didn't seem possible that this enormous piece of machinery could revolve. It had to be revolved by a hydraulic motor belt in the aircraft and the belt drive. The idea of this thing whizzing round in the aircraft and then being released just didn't seem feasible at the time. The bomb itself was like a big cylinder – a big plain cylinder. But we had no practice bombs in that shape – the only practice bombs we used before we got the special aircraft for the raid were of the ordinary lightweight explosive type. We never made any practices with the actual weapon, and never even saw it until the night before we took off.

Flight Sergeant George Chalmers

Wireless operator, AJ-O

The mine itself was just like a big oil drum – but a bit bigger, possibly larger in diameter, not much more. It looked like an oil drum made of metal, probably about half an inch thick metal all the way round.

Harold Roddis

Flight mechanic

We only saw the bombs when the armourers came out with them before the raid. The same wag who called the Lancs 'Abortions', thought that after being bombed up, they looked like pregnant ducks.

Flight Sergeant William Townsend

Pilot, AJ-O

We never bombed with the actual type of bomb that we used on the dams raid – they were in short supply. In fact we only had one practice with that bomb and that was down on the south coast, in Manston. We flew out to sea and then flew in towards the coast on which were erected two small towers, obviously to simulate the silhouette of the dams – and we just attacked these and rolled the bomb from the aircraft, bouncing it on the water on to the coast and inland. That was the only practice we had with the bomb.

On this particular trip we had Les Munro as our pilot, and it seems that we dropped it when we were a little bit below the correct height. It being daylight, we were just guessing the height, 50 feet, and the thing just bounced off the water and hit our tail. It didn't bring us down, but it was a very uncomfortable experience – and I think it damaged the rear turret. The aircraft was shaking quite a lot – which I hadn't expected – and in fact when I went on the raid my aircraft shook even more. It was vibrating very, very badly.

This was caused by the bomb rotating. It was rotating at 500 rpm, and when you've got that hung outside in the slipstream it's going to do something to the aerodynamics of the aircraft – it's going to shake it up.

Flight Lieutenant Les Munro

PILOT, AJ-W

The airspeed would affect the rate of trajectory of the weapon, and it could have been a combination of that and being a bit lower than the 60 feet required. My rear gunner, Flight Sergeant Weeks, was jammed in his turret because of the damage caused by the splash. I felt the thump, but there was no danger of going in the drink, even though I didn't climb away as we had to stay at low level for the training.

Wing Commander Guy Gibson

Next night we carried out another dress rehearsal, and it was a complete success. Everything ran smoothly and there was no hitch: that is, no hitch except that six out of the twelve aircraft were very seriously damaged by the great columns of water sent up when their mines splashed in. They had been flying slightly too low. Most of the damage was around the tails of the aircraft: elevators were smashed like plywood, turrets were knocked in, fins were bent. It was a miracle some of them got home. This was one of the many snags that the boys had to face while training. On the actual show it wouldn't matter so much because once the mines had dropped, the job would be done and the next thing would be to get out of it, no matter how badly the aircraft were damaged by water or anything else. But the main thing was to get the mines right on the spot.

By now it was obvious that we would have to carry out the raid within the next few days – perhaps only two days, because the water level had been reported just right for the attack. The training was now complete. The crews were ready. In all we had done 2,000 hours of flying and had dropped 2,500 practice bombs, and all the boys were rather like a team of racehorses, standing in the paddock waiting for the big event. But the ground crews were working like slaves repairing the damage to our aircraft.

Flight Lieutenant David Shannon

Pilot, AJ-L

We didn't get the mines we were going to drop until three days before the raids. We actually saw the mines coming in by road transport on low-loaders, but covered under tarpaulins. They were taken into the bomb dump and there was utmost secrecy there, so we never saw them until they were brought out.

They looked rather like a huge oil-drum – a most cumbersome-looking thing. They were carried athwartships, so that you got the revolving of the drum on the axle which went through the centre of the bomb. Two arms camp up, opened out and clamped into it, and to release the thing you lifted the arms away and it dropped straight down. It had the speed of the aircraft with its forward momentum, and the spin to give it the skip so that it didn't bury itself in the water. The object was to get a reverse spin on the mine, up to 500 revolutions per minute, driven by a belt run from an electric motor in the fuselage – which created a tremendous vibration in the plane.

A number of us were given the opportunity to go down and drop a dummy over the sea off Reculver – near Margate – to get the feel of how the aircraft would react when the 9,000 lb of revolving steel were released. The vibration was the thing – you had to try and hold the plane steady.

In the early days, all the practice bombing we'd done was just dive-bombing on a target, releasing by sight and judgement, using practice bombs over Wainfleet. But now we were told we had to line up on a couple of posts on the beach. We had to fly towards them just to get down to the height and line up with them with a handheld 'V' bombsight, which would determine the distance we were away before we dropped the mine. When released the mine behaved like a flat stone that you skimmed across the water. The aim was that it would settle up against the dam wall, then it would run down the dam wall and explode with a hydrostatic fuse some 30 feet under the water. With the weight of the water behind it, Barnes Wallis had calculated that one of these things should shatter the dams.

As far as we were concerned, the practice was quite successful. The dummy bombs were filled, I think, with concrete to simulate the exact weight – something between six and seven thousand pounds, fully armed. It had a hefty steel casing to stop it breaking. Even so, if you dropped it from too high, the casing would shatter. This practice allowed us to get the feel of the aircraft loaded at such low level with this thing spinning inside it, and also to understand the behaviour of the aircraft when you released it.

Sergeant Ray Grayston

Flight engineer, AJ-N

When we first saw Upkeep we were amazed at the idea that it would work at all. It horrified us to think they'd put this bloody great lump of metal, like a roller, underneath the Lancaster, which interfered so much with the slipstream – it interfered very much with the handling of the aircraft. The other problem we had with it was the spinning. When we turned it on, you'd turn on the hydraulics directly controlling this – you had to watch your aircraft hydraulics, so that they didn't lose all their oil. The first time we flew with it we lost all our hydraulics, but we didn't have the mine on board, as it was a dummy run.

When we did turn it on, it was rotating at 500 rpm, and the vibration was such that the instruments were nearly unreadable. Other than that it worked.

Flying Officer Harold Hobday

Navigator, AJ-N

The mines we were going to use weren't made until right before the raid, so we didn't do any training with the mine attached. They had lots of troubles with the manufacture of them, so we didn't have them until right at the end. We had a dummy mine on when we did our target practice over Reculver – but we only practised twice with the actual bomb on board.

Flight Sergeant Leonard Sumpter

Bomb-aimer, AJ-L

It was two days before the dams raid, and we'd got inert bombs. We took them down to Reculver beach. We had to sight on the Reculver church towers (as if they were the dam's towers), and release a bomb running into the beach. I released mine about forty yards short, so that it didn't make the distance. Gibson had me in the next morning and told me off about it. He was quite all right. He was a bit of a martinet, but he was a jolly good chap, discipline-wise. He didn't accuse me of being slack – but he said I wasn't paying attention. The pilot was in with me – Dave Shannon – and he put in a word for me because we got on well. We were well up in the bombing table, so it was just a matter of telling me to watch it and not do it again. There wasn't much time to do it again, anyway, because the dams raid was two days later.

I suppose it might have been a lapse in concentration, because when you see so many things in front of you through your sights, your eyes tend to wander a little bit. And this was daylight and there were towers there, and houses, and people on the beach – and these things were a bit distracting. But I wasn't the only one – another chap overshot with his and it went up on the promenade – a couple of photographers who were there had to dive out of the way, because it skewed round a bit.

The authorities had made films of the bomb bouncing – but we never saw those before the briefing. Some four or five crews got to see the bomb in use when we went to Reculver to test the equipment – that was the first time I saw the bomb bouncing on water. We thought it was marvellous. It was hard to think, when you first saw it, that when about five tons of metal hits the water, whether it's spinning in a backwards direction at about 450 revs a minute will make any difference to it. You'd think it would just go straight down – but it didn't. It just hit the water and up it came – and did about three bounces to the beach. It was very impressive – mine did four bounces to the beach – which was why I got chewed off. I dropped it too soon and it would have hardly reached the dam wall. The idea was to bounce it up to the dam wall and with it spinning against the wall, sticking to it through this spinning action, it would sink to the depth where the hydrostatic fuse would detonate it.

We didn't get the bombs until about two days before the raid, and a lot of them didn't arrive until the very day. I don't think they were even on the station – if they were, nobody knew about it. The bomb looked like a big olive green tar barrel with black ends, about four or five feet in length. At each end there was a ball-bearing swivel, and a hole in that fitted into the arms below the plane which clamped it in and held it tight – then it could revolve freely.

IWM FLM2343.

An inert Upkeep is dropped at Reculver. Barnes Wallis, extreme left, watches as his bouncing bomb theory is put into practice.

On one side of it there was a drive wheel, of the sort you used to have on the old motor bikes when they were driven by a rubber belt. This drive wheel and the rubber belt led to a wheel in the side of the plane, which led to a box operated by the wireless operator. This was driven hydraulically from one of the engines, so that when he opened the valve it started the mechanism going, turning the bomb until it reached the correct speed. This had to be done ten minutes before you were going to be called in to bomb. So when you got to the target, your bomb was spinning.

This meant the pilot was flying a gyroscopic aeroplane – because this had a gyroscopic effect, spinning underneath at 500 revs in a backwards direction. This made it a little harder for the pilot to steer the plane. It was up to the wireless operator to let the pilot know when it was spinning at the right speed. It would be the navigator's responsibility to line up the two spotlights on the water – the one in the noise pointing to the rear and the one about halfway down the plane pointing to the front. These were offset so that the two beams shining on the water gave you a white circle. When they were together, side by side, that gave you 60 feet. If they had crossed over, you were too high – and if they hadn't met you were too low.

In the meantime, you'd be at the front, with your string, watching the towers come up, listening to the navigator correcting the height, waiting for, 'Height's OK'. You'd ask the pilot, 'OK?' and he'd say, 'All right, bomb in ready' and you'd drop your bomb. If things weren't right, he'd say 'Abort' and we'd go round again.

Flying Officer Harold Hobday

NAVIGATOR, AJ-N

To pick out the first nine crews, it was decided that we'd have a dummy run with a dummy bouncing bomb at Reculver in Kent, and the crews who got the nearest to the target would be picked in order of seniority. Fortunately we got pretty near the target, so we were picked as part of the main force.

There was one fellow who was most upset because he wasn't in the main force – he was one of the pilots. He wanted Gibson to take us off and put him on, but Gibson wouldn't do it. That's the sort of man Gibson was – even though this man was a friend of his. Consequently, we went on the main force – and a very good job we did, too.

Wing Commander Guy Gibson

I realised the whole squadron was beginning to get a bit tired. They had had nearly two months of continuous training, getting up at dawn and dusk so that they could fly in conditions which were nearly like moonlight. They had flown about a hundred hours of training, and because of the uncertainty of the whole thing, the strain was beginning to get them down. For this reason I sent them off on three days' leave so that they could recover, but told them not to say a word to anyone outside. And I began to get ill too, and irritable and bad-tempered, and, of all things, there began to grow on my face a large carbuncle.

I went to the doctor. He was very kind and he said: 'This just means you are overworked. You will have to take two weeks off.' I just laughed in his face. Poor man! These medicos sometimes do not realise that there is a war on and when a job's got to be done, personal health doesn't enter into the question. Nevertheless, I took part of his advice and began taking some tonic.

Squadron team spirit

Despite long hours' training and the stress of such demanding flying, the crews of 617 Squadron still managed to enjoy their free time. Lasting friendships were forged as they bonded through their teamwork – and with the threat ever present of sudden death, they were hell bent on having some fun when the chance arose.

Flying Officer Harold Hobday

NAVIGATOR, AJ-N

Relations among the crew were marvellous. Originally Johnson and I were the only commissioned people, then Les Knight, the pilot, was commissioned, but the rest were sergeants, so we didn't have much to do with them in our leisure time, because we were in different messes. During our flying time and working, we were all very friendly indeed – and a jolly good job we were. We relied on each other. We were great pals, and have always remained so. Certainly, any personal antagonisms would have disrupted the crew's work – you couldn't have it. No-one resented anybody else. Just because I was an officer and some of the others were not, no-one resented it.

Flight Sergeant Leonard Sumpter

BOMB-AIMER, AJ-L

On some squadrons the crews would go down to the flights to do their operation and might be off for two or three nights and they're all matey together – all the crews. But on 617 I found that with all the training flights – two a day from when you were formed, right up to the raid – you didn't get to know anybody else. You didn't get friendly with the other crews. You said 'Good morning' to them, yes, but you never got really intimate with them. You were your own little band of seven – you kept yourselves to yourselves because you were so interested in what you were doing.

With 617 you were at low level all the time and you'd got to keep your brain on the job. You couldn't afford to relax or think of anything else. That's why we were so thick.

Flying Officer Harold Hobday

NAVIGATOR, AJ-N

We had a bit of leisure time, and we used to go into Lincoln a lot, to dances, for instance, or we played snooker in the mess. There were no restrictions on our leisure, but we were put on our honour about security. Naturally we couldn't hop down to London for the night – but we had mess parties occasionally and asked people from other messes and stations to come over – we used to enjoy ourselves.

Flight Sergeant Leonard Sumpter

BOMB-AIMER, AJ-L

In our crew, the pilot was Flight Lieutenant Shannon, the navigator was Flying Officer Walker – a Canadian – the engineer was Flight Sergeant Bob Henderson, the wireless operator was Brian Goodale – who we always called 'Concave' because he was always bending forwards – not just when working, but when he was walking too. He was skinny with it, so we called him 'Concave'. He didn't mind – he thought it was a big joke. He was a great fellow and liked a pint, Big Brian. Then our rear gunner was Jack Buckley, and the front gunner was Brian Jagger.

There were some real characters – Jack Buckley had a racing car and always had a pint in his hand. Danny was the quiet one – the Canadian. Brian Goodale was a bit of a character as far as drinks were concerned – and there was Shannon – who was only twenty – but he like a pint now and again. I'd seen him a bit helpless at times. Then Lofty Henderson was a tall, staid Scotsman, and never had much to say for himself – he used to come out for a drink with us now and again, and we'd go out for a dance in Lincoln. Brian Jagger was the grandson of the royal portrait painter.

We all got on like a house on fire – never any bad blood at all. You couldn't afford to have bad blood when you were flying

Courtesy of www.ww2images.com.

Relaxing at the officers' mess, Scampton. Left to right, sitting: Joe McCarthy, Les Munro, 'Spam' Spafford, Dave Shannon, Guy Gibson and Ann Fowler (who was soon to marry Shannon). Standing: Geoff Rice, Richard MacFarlane and Richard Trevor-Roper.

together nearly all day long. I think all the crews on the squadron were the same. You'd do three hours and go and have your lunch, then you'd do another three hours, have some supper. Then perhaps you'd have another two or three hours at night. Sometimes you were flying three times a day if the ground crew could keep the plane going that long. If you had any faults on the plane you had to wait until they were fixed or borrow another one – but we didn't like borrowing planes because you got used to your own. It was like going into someone else's house – a stranger's house – you didn't feel right.

John Elliott

NCO GROUND CREW

The relationship between the ground crews and the aircrews was excellent. We all got on together, and I don't remember there being any problems. The aircrew and ground crew NCOs used to be in the same mess and did a lot of socialising. The same was so to a lesser extent with officer crew, as they were billeted separately at their own mess. It was generally NCOs in the NCOs' mess and officers in their mess. But outside the mess, well, we all used to mix.

Flight Sergeant Chalmers

WIRELESS OPERATOR, AJ-O

We were enjoying the war with all this low flying. There were quips being made by the other aircrew at Scampton. They were disgusted at just coming back from an operation, only to find we'd been flying around the country seemingly enjoying ourselves.

Flight Sergeant Ken Brown

PILOT, AJ-F

The other squadron at Scampton called us the 'Armchair Boys' – sitting in armchairs doing damn all.

Flying Officer Edward Johnson

BOMB-AIMER, AJ-N

We didn't speculate what the role was to be at the start – but we did after we'd done some training and there was some indication of where we were going. We made some guesses as to our target – and most of them were more terrifying than what it turned out to be in the end. One favourite was that we might be going to try to toss bombs into the entrance of U-boat pens and blow them up internally, since they hadn't succeeded in doing much damage externally. That sounded a pretty terrifying operation, since most of us had been to one or other of the Atlantic ports. We were rather relieved when it turned out to be the dams – which nobody knew much about.

Flying Officer Harold Hobday

NAVIGATOR, AJ-N

We'd been flying over lakes for a long time in Wales and all over the country, and flying down the lochs in Scotland, and we seriously thought we were going after the *Tirpitz*. We thought, 'They'll be bristling with heavy guns, light guns and everything else – we won't stand a dog's chance.' We were quite relieved when we found it was not the *Tirpitz*. However, we were used to being shot at – though mostly from about 20,000 feet where we used to fly – and we'd have shells coming at us from the anti-aircraft in Germany – and fighters too. We'd seen many planes shot down in raids previously, so we were not blasé, but fatalistic about it. I don't think any of us expected to get through really, because so many of our friends had been shot down or missing – or simply had accidents. The percentage was very high. I'm surprised I got through. All the same, I don't think people took special precautions – making wills – I certainly didn't.

Wing Commander Guy Gibson

The AOC came over on 15 May and told me that we would take off the next night. As he left, a white aeroplane landed on our aerodrome and taxied up to our hangar. Out of it stepped Jeff and Mutt. They had come to see the end of all their work.

Hoppy was now told the target, as were Dinghy and Henry as well as Bob. And that night Hoppy saved our lives. The route the SASO and I had planned between us had taken us over Huls – a vast, well-defended rubber factory on the edge of the Ruhr which wasn't marked on the flak map. He happened to know all this, and suggested that we should go a little farther north. If Hoppy hadn't known, we would have gone over that factory, and it would have been just too bad. We went to bed at midnight. As we were leaving, Charles came rushing up with a white face. 'Look here, Guy, I'm awfully sorry. Nigger's had it. He's just been run over by a car outside the camp. He was killed instantaneously.'

So I went back to my room on the eve of this adventure with my dog gone – the squadron mascot.

CHAPTER 5

The Briefing

Since the end of March, the squadron had responded to every new requirement stipulated by the boffins and planners. They had trained in low flying, night navigation and low-level bomb-aiming – and a few had had an opportunity to practise with the new bomb – but at last, on 16 May 1943, they learned what their target was to be.

Wing Commander Guy Gibson

It was a great moment when the public address system on the station said 'All crews of No 617 Squadron report to the briefing room immediately.' The boys came in hushed, having waited two and a half months to hear what it was they were going to attack. There were about 133 young men in that room, rather tousled and a little scruffy, and perhaps a little old-looking in spite of their youth. But now they were experts, beautifully trained, and each one of them knew his job as well as any man had ever known any job he was to do.

Flight Lieutenant David Shannon

Pilot, AJ-L

The briefing finally began on 16 May. In the first instance it was purely the pilots and the navigators who were called forward, because contour-scale models of the Möhne and the Sorpe dams – the prime targets for attack – were unveiled, and we were told to study them and get as much from the models as we possibly could.

Wing Commander Guy Gibson

It was a boiling hot day, but all aircrews spent three hours of the afternoon studying the models of the dams. They studied and stared at them and drew them and photographed them on their minds until they knew exactly what they looked like. Then they asked each other questions from memory as to the position of the guns, the length of water and other vital points.

As the day cooled down we had our briefing. I will never forget that briefing. Two service policemen stood outside on guard. The doors were locked. No-one was allowed in except the boys who were going to do the job and four other men. No-one, not even Fighter Command or our own Group Headquarters, knew we were operating. They thought it was just a training operation.

Flight Lieutenant David Shannon

Pilot, AJ-L

The Möhne Dam, near Soest, about 25 miles east of the Ruhr district, was the first target. It served all of the Ruhr industries with a capacity of 134 million tons of water. Once that one was breached, the aircraft would fly on to the Eder Dam, which was another 70 miles further east. This was serving all the hydroelectric systems for Kassel and its heavy industrial areas – then if there was anything

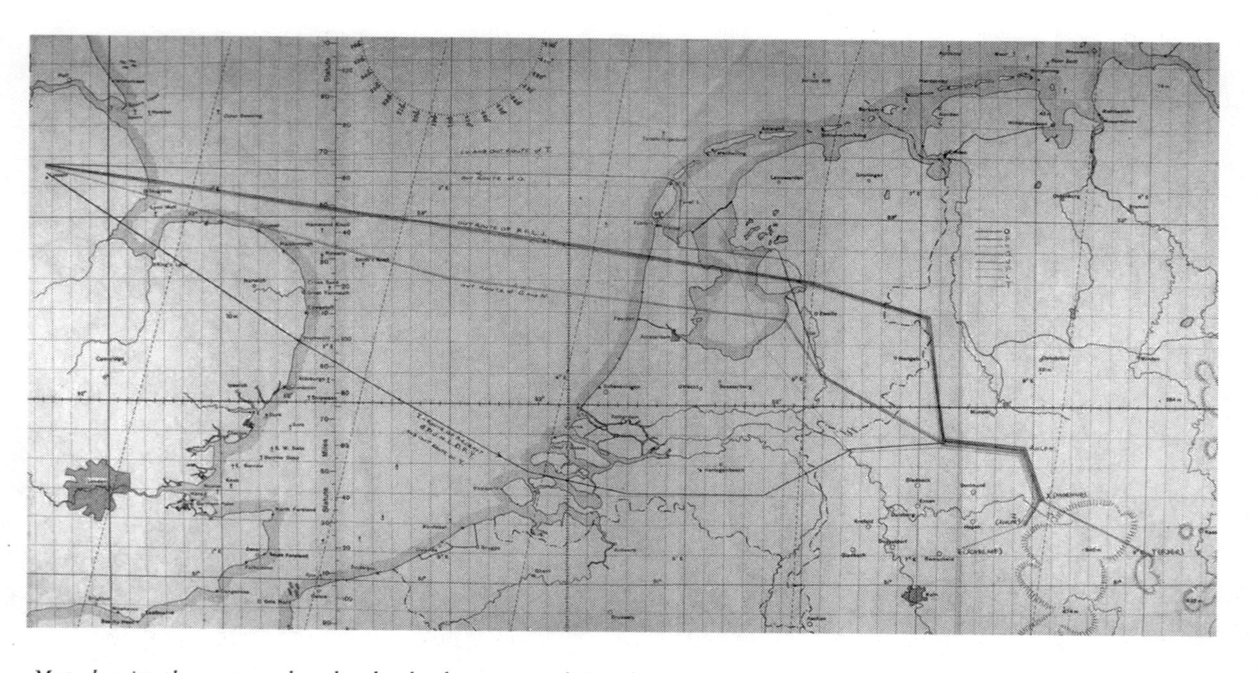

Map showing the routes to be taken by the three waves of aircraft on the night of 16/17 May 1943. Courtesy of The National Archive, ref. AIR14/840.

left, we'd go on to the Sorpe Dam. We also learned what the strategic significance of the dams was, and what our flight routes were. Gibson and the nav officer, together with some of the pilots with experience of Germany had mapped out the routes we would fly in, from Scampton across the North Sea, to some islands just off the Dutch coast, both of which were heavily defended, crossing the Dutch coast then into Germany. We'd then go slightly north along the Ruhr Valley, keeping over the Rhine – also a very heavily defended area, and then turn to the Möhne Dam first. We spent some two or three hours over this in the morning.

Then we were sent back and sworn to secrecy before the full briefing in the afternoon, when all the crews came in. Only then did they see for the first time what their targets were. There was a full briefing given by Gibson and the station commander and Group Intelligence.

Then Barnes Wallis explained his bouncing bomb. He told us the speed required at the time of release was 220 mph, at the height of 60 feet. The distance at right angles over the water back from the dam wall was 410 yards. The bomb would then bounce three times, arrive at the dam, roll down to a depth of 30 feet and then explode. There was also the normal intelligence briefing on enemy defences. We had had to wait for this time because the object was to hit these dams and breach them when they were absolutely full of water – to their maximum capacity – so as to do the most possible damage. Unbeknown to us, reconnaissance aircraft had been flying over the dams for the previous two months, watching them as the water level crept up. Also we had to wait for the full moon, because the only way we could operate at such low level was at a full moon – mid-May. The weather forecast was good for that evening. Security was very tight: no telephoning out, all mail was censored.

Flight Sergeant Leonard Sumpter

BOMB-AIMER, AJ-L

All the crews were at the eventual briefing – the one the day before had just been for the pilots and navigators, when they were told what the target was and shown the models of the dams. When it came to it, though, we never saw a model of the Eder – only the Möhne – so we didn't know what it looked like.

So, we were told the target and shown the model, told the route and what to watch for, and where to watch for flak – and details of what would happen when we got there. We would be called in one by one – up to nine, as there were nine of us going to the Möhne and Eder. The rest were directed to different dams – but they wouldn't take off until some two hours after us, at about half past eleven.

We knew that the target was going to be over water, but we thought it was going to be submarine pens. We had no idea that we were going to breach the dams – so we were quite taken aback when we heard what it was. I suppose we should have realised because of all the training we did over the Derwent Dam and the Eyebrook reservoir at Uppingham. In fact, they even put dummy towers on the Eye reservoir – there were already towers on the Derwent.

These dams were an unknown quantity to us – we knew it was the first time it had been attempted. When you went to Berlin, if you'd been there once before you knew what you were going to do. But these dams, being a one-off, you didn't know what to expect.

Gibson picked the crews for the Möhne with the best pilots, whom he thought could get it right. As far as I can see, the nine that went to the Möhne and Eder were his best pilots – not to say the others weren't good – they all were – but he just picked his crews.

Sergeant Ray Grayston

FLIGHT ENGINEER, AJ-N

It's not generally known, but Gibson, with what he had available, decided to use the more experienced crews and those with the greater ability on the essential targets – so as a result of that, we were number 9 in the pecking order, and we were on targets A and B – the Möhne and Eder dams.

John Elliott

NCO GROUND CREW

The ground staff didn't know what was on until the actual evening of this raid, and neither did we know what the target was going to be. Air Vice-Marshal Ralph Cochrane came down to our dispersal hut on the night that they took off and said, 'I realise how difficult it has been for you and the work you have had to put in for all of this, and I just hope that in the morning you'll consider it has all been well worth while.' But we still didn't know what they were going to do.

Flying Officer Edward Johnson

BOMB-AIMER, AJ-N

We only learned the purpose of our raid on the afternoon before – that was 16 May. Barnes Wallis, the inventor of the bomb was there, along with the CO, Gibson, Cochrane, and a couple of assistants with him as well.

They had models of the Möhne and Sorpe dams and they drew on blackboards, as they did at all briefings, showing the construction of the dams so we could appreciate what was going to happen when the bomb was released. We were then able to look at the models after the briefing to study the terrain and try to identify what we were looking for when we got there. We also had maps to prepare for the route because until then we didn't know the route we were flying on. Most people adopted a roller-towel technique of cutting up a map out of successive maps and mounting on a roller so that you could unfold it a bit at a time – rather than having a lot of maps all open in the bomb-aimer's compartment, where you're not blessed with all that much room. The tracks were marked on that – and perhaps the distances, ten or fifteen miles each side of the track and no more. This saved you from looking at things you didn't want to see. This map-reading is quite tricky at low level when you're whizzing over the ground – features you might see from high up are absolutely no good to you when you're low down.

Almost everybody spoke at the briefing – the AOC and Gibson, and Barnes Wallis, and the weatherman – a meteorologist. And the wireless operator would be there to say what the wireless drill was to be for the wireless operators. There was a little briefing on the flak situation on the routes we were to take. Not everybody was flying the same route because it was split up into several sections of operation. We were in the main wave – the one that went to the Möhne Dam. If we managed to crack that, we were to fly on to the Eder and attack that.

We could ask questions if there was anything important we couldn't understand – but there were not many questions. I think we were all a bit overawed at what the target was – it came straight out of the blue. Nobody had the slightest inkling before, and it took them a bit of time to digest what was going on. But it was all hubbub after the briefing was over.

After that most people were giving a sigh of relief that it didn't seem as dangerous as some of the other targets we'd been thinking about. It was short notice, but I don't think there was anything lost by that. The training we'd done was totally adequate to cope with what we were given at the briefing – and I don't think anything would have been gained by briefing us earlier. It would certainly have been bad for security, because there's always a chance, if it's out, that somebody will talk about it.

The briefing was in the afternoon and we took off around seven in the evening, so we had perhaps four hours to sort ourselves out and make maps, talk to each other if we wanted to discuss tactics, or what we might do in the event of particular contingencies. We'd plenty of time for that – no rush or panic.

I worked very closely with the navigator, Harold Hobday – we were very close friends, he and I, so we always got on and understood one another very well. Being trained as a navigator and observer meant I understood his problems as well as my own.

Before we took off, if anything we were slightly more elated than usual, knowing it was a very special target that had never been attacked before, and the fact that we were pioneering a low-level attack.

Flight Sergeant George Chalmers

Wireless operator, AJ-O

I turned around to my colleague and said, 'Who's the bloody civilian up there?' and he said, 'That's Barnes Wallis.'

'Barnes Wallis? What does he do?'

'You'll find out when the briefing goes on.'

So I sat down, and the briefing started, and we began to know what was going on.

Flying Officer Harold Hobday

Navigator, AJ-N

Barnes Wallis came along to the briefing and explained the functions and how the bomb was going to work. He described the action of the bomb, skipping along and dropping down just at the parapet – then exploding at a depth of 30 feet – and hopefully the explosion would crack the parapet of the dam itself. He was very enthusiastic about it, and explained it in minute detail, using a blackboard. He drew on it rather than using slides – but we had models of the two dams, and we studied those minutely to try and see what we were up against as regards terrain. On the Eder Dam it was particularly bad because there were hills on both sides which we had to scale, then drop down, and then we had to flatten out to 60 feet. Then as soon as the bomb had gone we had to climb to a thousand feet or so – it was very minute timing needed to get to the right place at the right time and the right height – and then get out of it alive without crashing. Barnes Wallis came across as a very kindly man – very dedicated, frightfully clever – but rather a fatherly type. We were very impressed with him and thought he was a marvellous man – everybody did.

We were absolutely astounded that he could have invented something which was unknown before. There had been nothing like it before, so it was just one of those innovations which one wouldn't expect. Who would expect a 9,250-lb bomb to bounce on water? I certainly wouldn't. But we weren't sceptical – we trusted him implicitly. We never dreamed that it wouldn't work. After all his efforts we didn't ever dream it would fail – and it didn't.

Sergeant Jim Clay

Bomb-aimer, AJ-W

It seemed incongruous that this kindly and quietly spoken man, Barnes Wallis, should be involved with devastation.

Flight Sergeant Grant McDonald

Rear gunner, AJ-F

When we went into the briefing room to hear what the target was there was a very tense atmosphere. It really didn't hit home when it was revealed. In those days you knew there was nothing you could do about it – you were going to go anyway. I was pretty confident I would come back.

Sergeant Frederick Sutherland

FRONT GUNNER, AJ-N

I think it was like going in to an exam room – you could feel the tension in the air. Everybody was keyed up.

Flying Officer Harold Hobday

NAVIGATOR, AJ-N

We aircrew were confident about the raid – we felt it was a marvellous operation to be on. It was so different from any bombing we'd ever done before, and much more exciting. And the planning was very smooth – first-class.

Gibson also spoke at the briefing, about our tactics, because as usual, it was explained exactly what was going to happen on this particular raid, as on any raid we'd been on. On this one, however, it was much more minutely directed, because flying low-level, we had to avoid, as much as we could, the German aerodromes and gun posts, high-tension cables, wires and other obstacles. We had what we called a 'dog-leg' system of navigation where we went a few miles one way and then a few miles another to avoid obstacles. We were also aided by the fact that a lot of the pinpoints, where we had to turn from one course to another on the dog-leg course were on canals, which were quite visible.

Sergeant George 'Johnny' Johnson

BOMB-AIMER, AJ-T

We only learned of the target on the actual day of the attack. We weren't briefed until the Sunday afternoon.

Our crew was briefed for the Sorpe Dam. The Sorpe was different in that it couldn't be attacked in the way that we had been practising. The geography of its situation was such that it had to be attacked by flying along the length of the dam.

Sergeant Dudley Heal

NAVIGATOR, AJ-F

We came to 15 May and we were called for briefing – the pilots, navigators and bomb-aimers, and we met Barnes Wallis and learned what the target was.

Obviously Barnes Wallis was a very clever man, and knew what he was talking about: the way the bomb was dropped, the height it was dropped from. He appreciated it was going to be quite a problem, and he explained about the need to spin the weapon. As it turned out, of course, we didn't do that, our crew, but I'll come to that in a moment. Anyway, we were briefed and sworn to secrecy and not allowed to leave the camp. And the next day, the 16th, all the crews were called and the final briefing was given, and we were told that we'd be taking off later that day.

We were absolutely amazed when we heard what the target was. It had certainly never occurred to me that we were going to attack a dam, because that was so different. I think of all the targets that were possible, the *Tirpitz* seemed most likely – and also the most dangerous – so I think we were quite relieved. At briefing we were shown models of the two dams – it was a very thorough briefing. We studied the target, we made notes. We studied the maps and everything was worked out in the greatest detail. It was then that we were told we would all be briefed to attack the Möhne and Eder dams to start with, and that we'd be given further orders en route depending on what success we had with those dams. We were one of the junior crews and we were in the third 'reserve' wave, which would take off some two hours after the first two waves and would go to whatever dam still needed breaching.

Flight Sergeant Leonard Sumpter

BOMB-AIMER, AJ-L

After the briefing you all go and mark all your tracks in. The navigator does his flight plan and you got ready to go with all your maps chinagraphed and with notes on the route. You'd mark in thick red crayon where the power lines crossed your track, because Dutch pylons were very high – a hundred feet or more. You had to watch out for those, but you'd have to go over or under them.

You'd do your flight planning and get your maps ready. You left all your kit in your bag in your locker – then you'd go back to your respective messes to have a wash and a shave, or a bath – whatever you felt like doing. No drinks, of course – just lemonade or something like that. Then you'd have your flying supper, which was always eggs and bacon or similar. If you didn't like eggs, you could have pancakes.

Then you'd go down to the flights about an hour and a half before take-off. You'd get your flying rations issued from your mess – usually two or three bars of chocolate, tin of orange juice, an orange, some chewing gum, Horlicks tablets if you liked them, and you'd pack these into your pockets and go down to the flights.

Then you were issued with your escape equipment – a wallet containing currency of the counties you were going to fly over, and a little compass, a long tin like a puncture repair kit with about three ampoules of morphine in it with needles to use. You could just take the cover off the needle and if anybody was wounded or in pain, you could squeeze the morphine out. Finally there was a silk map of the countries you were flying over with all the boundaries on, towns, main roads and railways, so that if you were able to, you could use this to escape to Spain or Denmark or somewhere neutral.

The training we did was spot on for the task in hand, and when we came to taking off, we were so used to conditions – except for having the bomb on board – that it was just like a training flight.

Flight Lieutenant David Shannon

PILOT, AJ-L

After the briefing we adjourned back to the mess, had a pre-operational meal and then waited for take-off.

I don't think one would talk about apprehension before the raid – every individual had his own feelings, but one had joined the forces to defeat the enemy, and apprehension in the form that we know it today didn't really occur. All the men that had been picked for this squadron were experienced bomber crews and had all completed at least one tour of operations – some of them more. All of them had volunteered because everybody was given a choice – it was not an order that one joined this squadron. Everyone had the opportunity to say yea or nay, and in fact, some of the crews that joined in the early days Gibson got rid of as unsatisfactory, and posted them elsewhere. No-one left voluntarily, to my knowledge. So feelings before this raid and after the briefing were of elation, excitement and, to a certain extent, relief that the training we'd been doing was coming to an end, and here we could see that the operation was about to take place. We were to strike a blow against the Hun. At last we were going to do something, because we had done nothing but practice. We were on an operational station with another bomber squadron operating at night there, and nobody knew what we were training for, and there was a certain amount of talk of 617 – 'the squadron does nothing but practice-fly – when are they going to do something?' So we were all very keen to get on and do it.

Looking back, the training that took place and the concentration required to achieve the accuracy required was only a period of six weeks – which is not really very long, although it seemed an age at the time.

Flight Lieutenant Les Munro

PILOT, AJ-W

I remember going into the ops room and looking at the maps and seeing the tracks leading to the dams. And then we realised what they were and the implications of that – that the route to the dams took us right up the Ruhr, one of the most heavily defended areas of Germany, that was when there was a certain amount of disquiet among the crews – but it was something we just had to accept. It was another operation we had to carry out and we had to do our best.

Fatalism – that was mostly my attitude throughout the war. Seeing the target at the briefing, to me I felt no different to if the target had been Berlin and all the consequence of having to go there and the defences – to me it was a question of accepting fate. If I was going to buy it, then so be it, but I never let that determine my feelings on the flights to the target. I had to concentrate on my job as pilot, getting the plane there and back.

Sergeant Ray Grayston

FLIGHT ENGINEER, AJ-N

We all normally went to the mess, and they gave us a cracking meal, because it might be your last one.

Flight Lieutenant Harry Humphries

ADJUTANT, 617 SQUADRON

I knew it was going to be something special and we'd already had the flying programme and two briefings, and they all gathered round the flight offices on the ground waiting for the transport to get to the aircraft. That really was something – you knew that something historical was going to happen.

Chapter 6

The Raid

The aircrews now knew their targets – and while there was some relief that they were not to take on the heavy defences surrounding the U-boat pens or the massive guns of the Tirpitz, *they were all too aware of the risks involved. For some, premonitions and doubts overshadowed their departure as the Lancasters took off in three waves. Oddly, the 'Second Wave' took to the sky first, as their route to the Sorpe Dam was plotted to be longer than that of the 'First Wave' which, led by Gibson, left Scampton in three wings of three, the last taking to the sky at 10.00 pm. The Third Wave, a mobile reserve, took off shortly after midnight, with a brief to follow the same route as Gibson's wave – and then await instructions. All would fly at low level throughout the mission to avoid detection by German radar.*

John Elliott

NCO GROUND CREW

All through that day I don't think any of us really knew that this was *the* day. The aircrew knew, yes, but the rest of us were not sure because it was much like any ordinary day. Everything that could be serviceable was serviceable, ready to go. It could have been a practice trip, but early in the evening was the first time we realised something was going to happen that night. Cochrane came down and told us.

Ruth Ive

WAAF, Radio operator

I used to be sent round the country to military camps and air force stations where there were to be special exercises or something special was going to occur, and where they wanted a clamp on security. We used to plug into the PBX – the telephone exchange – that served the camps and the air force.

I was told to go up to Lincoln, and I was met at the station by an RAF sergeant who said there was going to be a very important raid that night. I was driven on the airfield to a Nissen hut at the end of the airfield where they had fixed up the equipment for intercepting telephone calls. One of my very minor accomplishments in those days was that I knew how to bug a telephone exchange. Very easy then, with a hairpin or a paperclip.

You had to disconnect a lot of wires, and you have a little junction box, with all the lines that you wanted to listen to wired in. Eventually these would be wired up into a small box which you could plug into – not unlike stereo loud speaker but with lots of switches to it. You took the lines from the adjutants' mess out of their junction box and put them into another one and they could still have the conversation with the outgoing person – although on this particular raid they weren't to be allowed to make telephone calls at all. So we knew if our lines were being used by a little light or flickering ball. Anyone making a call would be going against the rules because there was a 24-hour clamp down on security and all telephone calls. They put a fix on the officers' mess and various other places, and I had to listen to the telephone calls and make a note of any indiscretions and disconnect the call at the same time. I still hadn't any idea why I was there.

Flight Lieutenant Harry Humphries

ADJUTANT, 617 SQUADRON

From the take-off and landing times, I had to work out such things as the buses to take the aircrews to their planes in good time to complete the preliminary crew drills and prevent unnecessary rush and a mealtime to allow the boys to catch the buses. In between mealtimes and bus-times I had to ensure that flying rations were available, and coffee which the aircrew carried in the aircraft while on operations. I had to be available to accept cash, wills and letters to next of kin, should anyone wish to deposit such articles. Strangely enough, the men who did deposit anything with me always seemed to return to collect.

I overheard the dialogue after the briefing and before take-off.

'I hope they all come back,' Wallis said.

'It won't be your fault if they don't,' Gibson replied.

Wing Commander Guy Gibson

Hoppy said to me, 'Hey Gibbo, if you don't come back, can I have your egg tomorrow?' I realised that the first aircraft to attack the dam will probably catch the gunners with their pants down – but the second to attack won't be so lucky – and that would be him.

Flight Lieutenant Harry Humphries

ADJUTANT, 617 SQUADRON

Evening came and all the squadron personnel, the aircrew, the motor transport drivers, the cooks, the mechanics, the armourers, the photographers, had either done their job of work or were standing

Wing Commander Gibson and crew boarding their aircraft on the night of the raid. Left to right: Trevor-Roper, Pulford, Deering, Spafford, Hutchison, Gibson and Taerum.

by ready. It was a beautiful night again, warm and inviting and in the messes the aircrew were making their final preparations. Those not on duty were on the airfield and waiting and watching for the take-off.

In the officers' mess, there were a lot of strange faces; high-ranking Air Force officers and several studious-looking civilians – the 'back-room boys'. I felt awfully small in the midst of all this activity. I wanted to help everybody to do something, but no help was needed. We had all done our little piece of the jigsaw and we were ready. I went to my office at six. And waited.

I managed to catch the eye of the Wing Commander and said, 'Anything you want me to do, sir?'

'Always on the spot, Adj,' he replied. 'I don't think there is anything at the moment – oh yes, there is though. Plenty of beer in the mess when we return. We'll be having a party.' He looked thoughtful and added, 'I hope.'

With that he turned to his crew and said, 'Well chaps, my watch says time to go. Cheerio Adj – and don't forget the beer.'

John Elliott

NCO GROUND CREW

I remember Gibson going. I was there with him and his crew, and they were having a mooch about the ground before they got in. He said to me, 'What's the time?' We had a look and he'd got the same time as I had, and he said to me, 'We'll start up at 9.25,' which we did and – away they went. Then it *was* just a question of waiting.

Flight Lieutenant Harry Humphries

ADJUTANT, 617 SQUADRON

The dramatic part is when it's silent, then they open the throttles and start the engines, and start waddling along to the take-off point

– that's the exciting part. They took off in formation, very, very low – and that was a tremendous sight. I knew it was going to be an historical night.

Ruth Ive

WAAF, Radio operator

Somebody brought us out some dinner, and then the whole place absolutely rattled and shook. The ground seemed to heave as the planes took off. And they just seemed to miss our hut. Wave after wave of them. I thought it was just a particular raid on Berlin, but the two sergeants said they weren't letting on where they were going. They said they didn't know – but we waited up all night – so they taught me to play poker.

Harry Humphries

Adjutant, 617 Squadron

I was adjutant of the squadron and I never knew what the target was until they came back!

I stayed until both lots had taken off, then it all went quiet. We got back to the mess – and we just waited. I sat in the mess next to a WAAF intelligence officer, Fay Gillon, and she said to me, 'Isn't this exciting, Humphy?'

Kenneth Lucas

Ground crew

Some of the planes were damaged two days before the raid – the tailplanes – I forget how many we had to replace – but this just had to be done, so I was one of the people who worked all day, all night, and until three o'clock on the day of the raid, because they just had to be ready to go. When they actually took off on the raid I was fast asleep in my bed.

Beck Parsons

GROUND CREW

On the night of the raid I was on duty – there was a duty electrician – but we didn't really know exactly what was going on. We knew it was big, but we didn't know until they took off what the target was, because the one thing about Gibson was that he was determined that nobody would leak it – he knew it was a big thing.

Sergeant Dudley Heal

NAVIGATOR, AJ-F

So on the evening of 16 May we stood and watched the first two waves take off with this enormous thing, almost like a garden roller, hanging underneath. We watched them leave in formations of three, lumbering across the aerodrome and over the perimeter fence and gradually climbing away. We thought – well, I'm sure the pilots thought particularly – 'I hope I can do that.' And then we had two hours to spend before we could take off. So we did our usual and went and played poker. I had already written a letter home – we always did before these missions. Then at about 11 o'clock that night we went down to our aircraft, F for Freddie, and spent the next hour or so with our ground crews, expressing our appreciation – we knew what wonderful blokes they were, how much trouble they took to make sure everything was right.

CHAPTER 7

The First Wave

The flight to the target

FLIGHT LIEUTENANT SHANNON'S CREW

Flight Lieutenant David Shannon

PILOT, AJ-L

In the end there were nineteen of us. As we were walking out to our aircraft Hoppy Hopgood grabbed me and we went round the back of a hangar to smoke a cigarette. He said, 'I think this is going to be a tough one, and I don't think I'm coming back, Dave.' That shook me a bit.

'Come off it, Hoppy,' I said, 'you'll beat these bastards; you've beaten them for so long, you're not going to get whipped tonight.'

At 9.30 pm with a full moon, we took off. We were the first wave; nine of us, in three formations of three, went with Gibson. Our target was the Möhne Dam, and if we broke that, we were to go on and do the Eder Dam.

Gibson and Martin and Hopgood went first. They were followed

IWM CH18006.

As dusk falls on the night of 16/17 May, a lone airman watches as a Lancaster takes off for the raid.

by me, Dinghy Young and David Maltby. Then came Henry Maudslay, Bill Astell and Les Knight. That was the first nine. At the same time our nine aircraft took off, immediately after that, five aircraft took off for the Sorpe Dam, flying on a slightly different route – so there was to be an attack concurrently on the Sorpe and Möhne dams. That got fourteen aircraft off at that stage, then there were another five aircraft held in reserve, to be sent off to other dams – because there were other dams as well, of which we didn't have models at the briefing, but there were photos. If we were successful with our first three, we could make further attacks on the other dams on the same night. But they were to be held back until such time as they had information, so they would make a very late take-off – about one in the morning – after the results had been radioed back of success or otherwise on the breaching of the Möhne and Eder dams.

Flight Sergeant Leonard Sumpter

Bomb-aimer, AJ-L

We took off in formation in threes and stayed low – at about 200 or 300 feet to Southwold, the turning point on the English coast. We stayed low out over the sea and across the little Dutch islands. They'd planned the route so that we followed canals, or had little lakes as turning points. I didn't see anybody shot down, because when you're down in the nose you can't see sideways. All you can see is ahead. The pilots of the other planes saw one or two of the chaps being shot down – and there was quite a bit of light flak on the route. But as bomb-aimer you don't see anything to port or starboard. So as far as I was concerned it was uneventful – until we got there.

En route, one chap, Astell, got off course – I think he was in our three, because we flew out in formation. He collided with high tension cables north of the Ruhr and that was him finished.

Flight Lieutenant David Shannon

Pilot, AJ-L

We formed up in our vics of three and flew across the North Sea – quite uneventful until we hit the coast of Holland, then the light flak opened up from there the whole way. The Dutch coast was very heavily defended and there was a lot of light flak. We got through, and then the main hazards going on were light flak, because, whereas the information we'd had of the placements of the light flak across the route was not entirely right, we did go through one of the new aerodromes that had been put in by the Germans, which was very heavily defended. I think it was there that the first of the aircraft was lost. It turned out to be Bill Astell. It was light flak and searchlights – but flying at that level, although the searchlights would come up, we were too low for them to catch us in the beams and hold us there for any length of time. But the recollection of the route in is of the tracer that one could see from the light flak and the searchlights waving. I believe there were plenty of fighter aircraft about, but we were down on the deck and the fighter aircraft didn't really know whether we were enemy aircraft or their own. Anyway, it was too low for the fighters to operate, so there were no attacks from fighter aircraft.

Flying Officer Daniel Walker

Navigator, AJ-L

When we were on our way in we were picked up by searchlights. They were dazzling. The blue light was a master light. Once the blue light was on you, the other ones automatically picked you up. But we were fortunate. We were able to shake it off. But we lost the other two aircraft we were going in with, so we just carried on singly.

Flight Lieutenant Leonard Sumpter

Bomb-aimer, AJ-L

We got to Soest, which we had to miss because there was quite a bit of flak there, then it was another 25 miles almost due south to the Möhne Dam, and we broke up and flew in individually. We came in over one of the towers and were sprayed with flak – which I didn't like very much – and the plane got two or three holes – but these weren't any trouble at all.

Pilot Officer Les Knight's Crew

Flying Officer Harold Hobday

Navigator, AJ-N

As we went out to our aircraft we had our battledress on. The trouble was that, being a navigator, I was lumbered with a sextant, which was quite a big thing, a navigational bag which is also quite big, full of maps, dividers, rulers and other equipment, and my parachute – so I had a lot to carry. The bomb-aimer had a special map he'd marked up – but the rest just had parachutes, their head-dress, oxygen mask and regular kit.

We weren't allowed to take anything personal in our pockets – that was very strict. The idea was, if you landed in German territory, if you had anything personal they would be able to tell where you came from, your squadron and so on – so you couldn't even take a wallet – or family photos. We never thought of that. We had a map, printed on silk, and some foreign currency in case we baled out, but that was a standard thing on all raids. If we were going over Holland and Germany, we'd have some Dutch money.

We took off and formated into threes. We were the last three taking off on the main force. I was navigating all the time, my head bowed down over maps a lot of the time over the North Sea to make absolutely sure that we were going to hit the coast at the right place – which we did. Once we got over to the other side, we were in threes and we started crossing Holland on our dog-leg course.

Sergeant Frederick Sutherland

FRONT GUNNER, AJ-N

There was a little shack on the edge of the runway rather than the control tower and there was a guy in there with a green light. The three of us lined up, and when we got the green light we revved up and all went down the runway together. It was a hell of a long run because we had so much weight. But once you get in the aircraft, no matter how scared you are before or how apprehensive you are, once you start down the runway, well, then it's OK. You're committed and what happens, happens.

Flying Officer Harold Hobday

NAVIGATOR, AJ-N

For navigating we had the GEE set – our radar navigation – which was very useful, and they very cunningly selected a special frequency for it which hadn't been used before, so the Germans couldn't jam it. That meant I got GEE coverage quite a long way out across Germany, which was quite unusual, especially at a low altitude. I was very lucky – although it didn't really navigate for us – it checked that we were on the right course. In the event, we lost GEE coverage just before the Möhne Dam.

After a couple of dog-legs I noticed that one of our three was going too much to the right, so we broke off from him and the next thing he went over what must have been an airfield, and he got shot down straight away. I knew who the crew were, and it brought it home to me – but one had to press on. We went on from dog-leg to dog-leg without any real trouble until we got to the Möhne Dam – we weren't shot at because we followed the right track – at least I think that's why it was. On one occasion we nearly hit some high-tension wires somewhere in Holland, but managed to go over the top of them.

Sergeant Frederick Sutherland

FRONT GUNNER, AJ-N

There was very bright moonlight, though your eyes were accustomed to seeing in the dark, especially rivers and fields. But you couldn't see any detail in the distance. There'd be a gun go off at the side of you which you hadn't seen. Maybe in daylight you could have seen some movement or a gun, but at night you can't. You can't get that detail, so they've got first shad, and that's where you know you haven't got much of a hope if they were any good. If they're accurate, well, you're gone. But once they shoot and they don't hit you, well, you've got hope.

Sergeant Ray Grayston

FLIGHT ENGINEER, AJ-N

In the main I think they succeeded in keeping us away from the trouble spots and hot flak points as best they could, but I think Gibson advised one or two to reroute on the way to avoid flak hotspots.

The main fear was the high-tension cables – there was a huge grid system in Germany. It was hoped that the people who plotted the route for you had avoided these situations – that you were not flying on a course to intercept them.

Bill Astell flew into an electricity pylon and he and his crew were all killed – that's how close we all were to sudden death – but he was formating on us.

Flying Officer Edward Johnson

BOMB-AIMER, AJ-N

The flight out to the Möhne Dam was fairly uneventful. We didn't seem to have any map-reading problems or navigation problems. We were astonished at the lack of German fighter cover – it was a bright moonlight night and somebody must have been aware that we were on the way. We had to get there and they don't miss much on radar, even at low level. I was surprised that the opposition wasn't greater than it was – certainly their fighters didn't turn up while we were at the Möhne Dam. We were circling there for quite some time in bright moonlight – even without radar we must have been very visible.

WING COMMANDER GUY GIBSON'S CREW

Wing Commander Guy Gibson

PILOT, AJ-G

We had been flying for about an hour and ten minutes in complete silence, each one busy with his thoughts, while the waves were slopping by a few feet below with monotonous regularity. The moon dancing in those waves had become almost a hypnotising crystal. As Terry spoke he jerked us into action. 'Five minutes to go to the Dutch coast, Skip.'

Pulford turned on the spotlights and told me to go down much lower – we were about 100 feet off the water. Jim Deering, in the front turret, began to swing it from either way, ready to deal with any flak ships which might be watching for mine-layers off the coast. Hutch sat in his wireless cabin, ready to send a flak warning to the rest of the boys who might run into trouble behind us. Trevor took off his Mae West and squeezed himself back into the rear turret. On either side the boys tucked their blunt-nose Lancs in even closer than before, while the crews inside them were probably doing the same sort of thing as my own. Someone began whistling nervously over the intercom. Someone else told him, 'Shut up.'

We were flying low – we were flying so low that more than once Spam yelled at me to pull up quickly to avoid high-tension wires and tall trees. Away on the right we could see the small town, its chimneys outlined against the night sky; we thought we saw someone flash us a 'V' but it may have been an innkeeper poking his head out of his bedroom window. The noise must have been terrific.

It didn't take Spam long to see where we were; now we were right on track, and Terry again gave the new course for the River Rhine. A few minutes later we crossed the German frontier, and Terry said, in his matter-of-fact way, 'We'll be at the target in an hour and a half. The next thing to see is the Rhine.'

But we did not all get through. One aircraft, P/O Rice, had already hit the sea, bounced up, lost both its outboard engines and its weapon, and had flown back on the inboard two. Les Munro had been hit by light flak a little later on, and his aircraft was so badly damaged that he was forced to return to base. I imagined the feelings of the crews of these aircraft who, after many weeks of intense practice and expectation, at the last moment, could only hobble home and land with nothing accomplished. I felt very sorry for them. This left sixteen aircraft going on; 112 men.

As we passed on into the Ruhr Valley we came to more and more trouble, for now we were in the outer light flak defences, and these were very active, but by weaving and jinking we were able to escape most of them. Time and again searchlights would pick us up, but we were flying very low and, although it may sound foolish and untrue when I say so, we avoided a great number of them by dodging behind the trees. Once we went over a brand-new aerodrome which was very heavily defended and which had not been marked on our combat charts. Immediately all three of us in front were picked up by the searchlights and held.

Suddenly Trevor, in the rear turret, began firing away, trying to scare them enough to turn out their lights, then he shouted that they had gone behind some tall trees. At the same time, Spam was yelling that he would soon be shaving himself by the tops of some corn in a field. Hutch immediately sent out a flak warning to all the boys behind so that they could avoid this unattractive area. On either side of me, Micky and Hoppy, who were a little higher, were flying along brightly illuminated; I could see their letters quite clearly, AJP and AJM, standing out like Broadway signs. Then a long string of tracer came from Hoppy's rear turret and I lost him in the momentary darkness as the searchlights popped out. One of the pilots, a grand Englishman from Derbyshire, was not so lucky. He was flying well out to the left. He got blinded in the searchlights and, for a second, lost control. His aircraft reared up like a stricken horse, plunged on to the deck and burst into flames; five seconds later his mine blew up with a tremendous explosion. Bill Astell had gone.

It was hard to say exactly how many guns there were, but tracers seemed to be coming from about five positions, probably making twelve guns in all. It was hard at first to tell the calibre of the shells, but after one of the boys had been hit, we were informed over the R/T that they were either 20 mm type or 37 mm, which as everyone knows, are nasty little things.

For a time there was a general bind on the subject of light flak, and the only man who didn't say anything about it was Hutch, because he could not see it, and because he never said anything about flak anyway. But this was not the time for talking. I called up each member of our formation and found, to my relief, that they had all arrived, except, of course, Bill Astell. Away to the south, Joe McCarthy had just begun his diversionary attack on the Sorpe. But not all of them had been able to get there. Both Byers and Barlow had been shot down by light flak after crossing the coast; these had been replaced by other aircraft of the rear formation. Bad luck, this being shot down after crossing the coast, because it could have happened to anybody; they must have been a mile or so off track and had got the hammer.

Attacking the Möhne Dam

Flight Lieutenant David Shannon

Pilot, AJ-L

Once we got to the dams, we could see them, silhouetted there in the moonlight. Gibson said he would do a reconnoitre, and we were all told to hold off in the hills three or four miles back from the dam itself, to await results. He made his attack through and it was then we found out that the Möhne Dam was very heavily defended with light flak on the dam wall itself and on the sides of the approach.

Flying Officer Harold Hobday

Navigator, AJ-N

We found the Möhne Dam straight away – it was a big lake, well lit up by the moon, and you couldn't miss it if your navigation was accurate – and I liked to think that mine was. It was very well laid out, and a colossal area of water – but I wasn't admiring the scenery as much as making sure everything was OK in the aircraft. I had a very good view if I looked out to the side – normally I was sitting facing a blank wall with instruments, but I could swivel round and look through a blister on the right side. That's how I was able to look at the lights when they were focused on the water to check our height.

When we got to the Möhne Dam, the other three were there,

and the bombing started with Gibson making the first run. As ordered, we circled round a discreet distance away to see what was going on and stand by in case we were needed.

Wing Commander Guy Gibson

As we came over the hill, we saw the Möhne Lake. Then we saw the dam itself. In that light it looked squat and heavy and unconquerable; it looked grey and solid in the moonlight as though it were part of the countryside itself and just as immovable. A structure like a battleship was showering out flak all along its length, but some came from a powerhouse below it and nearby. There were no searchlights. It was light flak, mostly green, yellow and red, and the colours of the tracer reflected upon the face of the water in the lake. The reflections on the dead calm of the black water made it seem there was twice as much as there really was. I scowled to myself as I remembered telling the boys an hour or so ago that they would probably only be the German equivalent of the Home Guard and in bed by the time we had arrived.

The boys dispersed to the pre-arranged hiding-spots in the hills, so that they should not be seen either from the ground or from the air, and we began to get into position for our approach. We circled wide and came around down moon, over the high hills at the eastern end of the lake. On straightening up we began to dive towards the flat, ominous water two miles away. Over the front turret was the dam, silhouetted against the haze of the Ruhr Valley. We could see the towers and the sluices – we could see everything. Spam, the bomb-aimer, said, 'Good show. This is wizard.' He had been a bit worried, as all bomb-aimers are, in case they cannot see their aiming points, but as we came in over the tall fir trees, his voice came up again rather quickly.

'You're going to hit them – you're going to hit those trees!'

Terry turned on the spotlights and began giving directions: 'Down

– down – down. Steady – steady.' We were then exactly 60 feet.

The German gunners had seen us coming – they could see us coming with our spotlights on from over two miles away. Now they opened up and their tracers began swirling towards us; some were even bouncing off the smooth surface of the lake. This was a horrible moment; we were being dragged along at four miles a minute, almost against our will, towards the things we were going to destroy. I think at that moment the boys did not want to go. I know I didn't want to go. I thought to myself, 'In another minute we shall all be dead – so what?'

By now we were a few hundred yards away, and I said quickly to Pulford, under my breath, 'Better leave the throttles open now, and stand by to pull me out of the seat if I get hit.' As I glanced at him I thought he looked a little glum on hearing this.

The Lancaster was really moving and I began looking through the special sight on my windscreen. Spam had his eyes glued to the bomb-sight in front, his hand on his button; a special mechanism on board had already begun to work so that the mine would drop, we hoped, in the right spot. Terry was still checking the height. Joe and Trev began to raise their guns. The flak could see us quite clearly now. It was not exactly an inferno – I have been through far worse flak fire than that, but we were very low. There was something sinister and slightly unnerving about the whole operation. My aircraft was so small and the dam was so large; it was thick and solid, and now it was angry. We skimmed along the surface of the lake, and as we went, my gunner was firing into the defences, and the defences were firing back with vigour, their shells whistling past us. For some reason we were not being hit.

'Left – little more left,' Spam said, 'steady – steady – steady – coming up.'

Of the next few seconds I remember only a series of kaleidoscopic incidents. The chatter from Joe's front guns pushing out tracers which bounced off the left-hand flak tower. Pulford crouching beside me. The smell of burnt cordite. The cold sweat underneath my oxygen mask. The tracers flashing past the windows – they all seemed the same colour now – and the inaccuracy of the gun positions near the power station; they were firing in the wrong direction. The closeness of the dam wall. Spam's exultant, 'Mine's gone!' Hutch's Very lights to blind the flak-gunners. The speed of the whole thing. Someone saying over the RT: 'Good show, leader. Nice work.'

Then it was all over, and at last we were out of range, and there came over us all, I think, an immense feeling of relief and confidence.

We could see with satisfaction that Spam had been good, and it had gone off in the right position. Then, as we came closer, we could see that the explosion of the mine had caused a great disturbance upon the surface of the lake and the water had become broken and furious, as though it were being lashed by a gale. At first we thought that the dam itself had broken, because great sheets of water were slopping over the top of the wall like a gigantic basin. This caused some delay, because our mines could only be dropped in calm water and we would have to wait until all became still again.

We waited about ten minutes, but it seemed hours to us. It must have seemed even longer to Hoppy, who was the next to attack. Meanwhile, all the fighters had now collected over our target. They knew our game by now, but we were flying too low for them. They couldn't see us and there were no attacks.

Flight Lieutenant David Shannon

PILOT, AJ-L

Gibson went in first and made an excellent run. A huge bloody spurt of water went up hundreds of feet, but the wall was still there. Hopgood, who was Gibson's number two at the time, went next. He had been hit coming over with light flak and was hit again in the petrol tanks, so his wing on the starboard side caught fire. I think the bomb-aimer was probably killed because he dropped his bomb late, and instead of skimming up and then sliding down the wall, it bounced and went over the wall on the hydroelectric plant beneath. He did his best to gain height and got his plane up to about 500 feet, then it burst into flames and exploded in mid-air. Hoppy's prediction had been proved right. Burcher and Fraser somehow managed to get out of that terrible smash and pulled their parachutes. They were taken prisoner-of-war.

Pilot Officer Anthony Burcher

Rear gunner, AJ-M

Hopgood had been hit himself – in the head – and blood was pouring out of the wound, but he yelled to the engineer, 'Carry on and don't worry! Get out, you bloody fool. If only I had another 300 feet – I can't get any more height.'

Wing Commander Guy Gibson

Then Hoppy began going down over the trees where I had come from a few moments before. We could see his spotlights quite clearly, slowly closing together as he ran across the water. We saw him approach. The flak, by now, had got an idea from which direction the attack was coming, and they let him have it. When he was about 100 yards away someone said hoarsely over the R/T, 'Hell! He's been hit.' M Mother was on fire – an unlucky shot had got him in one of the inboard petrol tanks and a log jet of flame was beginning to stream out. I saw him drop his mine, but his bomb-aimer must have been wounded, because it fell straight on to the powerhouse on the other side of the dam. But Hoppy staggered on, trying to gain altitude so that his crew could bale out. When he had got up to about 500 feet there was a livid flash in the sky and one wing fell off. His aircraft disintegrated and fell to the ground in cascading, flaming fragments. There it began to burn quite gently and rather sinisterly in a field some three miles beyond the dam. Someone said, 'Poor old Hoppy.' And another said, 'We'll get those bastards for this.'

A furious rage surged up inside my own crew, and Trevor said, 'Let's go in and murder those gunners.' As he spoke, Hoppy's mine went up. It went up behind the powerhouse with a tremendous yellow explosion and left in the air a great ball of black smoke.

Flight Lieutenant David Shannon

Pilot, AJ-L

The dam was still standing, so the next run of that flight was Micky Martin. While he was doing the run Gibson flew down on one side to try and attract the flak that was coming up all along the wall and on the pillars.

Wing Commander Guy Gibson

Bob Hay, Micky Martin's bomb-aimer, did a good job, and his mine dropped in exactly the right place. There was again a gigantic explosion as the whole surface of the lake shook, then spewed forth its cascade of water. Micky was all right – he got through, but he had been hit several times and one wing-tank lost all its petrol. I could see the vicious tracer from his rear gunner giving one gun position a hail of bullets as he swept over. Then he called up, 'OK. Attack completed.'

Flight Lieutenant David Shannon

Pilot, AJ-L

All this time we were just concentrating on what we were doing – once you're flying over enemy territory you don't have any feelings except waiting your turn to go in and bomb. There was a certain amount of disappointment at this time – after all, we'd been told that one of these bombs placed in the right position, would blow up the wall. One could see that there was flak there – but we'd all seen flak before. There was nothing new about that.

So the next vic was called in and Squadron Leader 'Dinghy' Young went in. He had Guy on one side and Martin on the other, flying down. He had a very successful run. We thought, 'Christ, that must break the bloody thing.'

Then Maltby did a perfect run in. Gibson and Martin were still on either side, but by this time the flak gunners were becoming a bit more subdued anyway. I think they saw the splashes and things coming at them. But still, that one didn't seem to break it. I was just starting my run in when suddenly the wall collapsed and Gibson yelled, 'It's gone! It's gone! For Christ's sake, David, hold off.' The wall had gone. I think it was probably Maltby and Young between them who'd smashed this bloody great wall with their bombs in the right place. It was a tremendous sight, we just saw the tail end, but the water was getting more and more and faster and faster – and it poured out of the lake, taking everything in front of it. After a few minutes it was just an avalanche of mud and water and stuff going down the hill.

Each time a bomb had been dropped, we'd sent a signal back to base saying, 'No go, no go!' 'Nigger' was the code word if we broke one, and after five runs on it, and the loss of an aircraft, we eventually got the 'Nigger'. Now Martin and Maltby were told to return to base. Young, as second in command, was told to go with Gibson and to control the bombing of the Eder Dam.

Flight Lieutenant David Maltby

PILOT, AJ-J

Our load sent up water and mud to a height of a thousand feet. The spout of water was silhouetted against the moon. It rose with tremendous speed and then gently fell back.

Wing Commander Guy Gibson

As Melvin Young came in next, we stayed at a fairly safe distance on the other side, firing with all guns at the defences, and the defences, like the stooges they were, firing back at us. We were both out of range of each other, but the ruse seemed to work, and we flicked on our identification lights to let them see us even more clearly. Melvin's mine went in, again in exactly the right spot, and this time a colossal wall of water swept right over the dam and kept on going. Melvin said, 'I think I've done it. I've broken it!' But we were in a better position to see than he, and it had not rolled down yet. We were getting pretty excited by now, and I screamed like a schoolboy over the R/T, 'Wizard show, Melvin, I think it'll go on the next one.'

When at last the water had all subsided, I called up No. 5 – David Maltby – and told him to attack. He came in fast, and I saw his mine fall within feet of the right spot; once again the flak, the explosion and the wall of water. But this time we were on the wrong side of the wall and could not see what had happened. We watched for about five minutes, and it was rather hard to see anything, for by now the air was full of spray from these explosions, which had settled like mist on our windscreens. Time was getting short, so I called up Dave Shannon and told him to come in.

As he turned I got close to the dam wall and then saw what had happened. It had rolled over, but I could not believe my eyes. I heard someone shout, 'I think she's gone! I think she's gone! Other voices took up the cry and quickly I said, 'Stand by until I make a recco.' I remembered that Dave was going into attack and told him to turn away and not to approach the target. We had a closer look. Now there was no doubt about it – there was a great breach 100 yards across, and the water, looking like stirred porridge in the moonlight, was gushing out and rolling into the Ruhr Valley towards the industrial centres of Germany's Third Reich.

It was the most amazing sight. The whole valley was beginning to fill with fog from the steam of the gushing water, and down in the foggy valley we saw cars speeding along the roads in front of this great wave of water, which was chasing them and going faster than they could ever hope to go. I saw their headlights burning and I saw water overtake them, wave by wave, and then the colour of the headlights underneath the water changing from light blue to green, from green to dark purple, until there was no longer anything except the water, bouncing down in great waves. The floods raced on, carrying with them as they went viaducts, railways, bridges and everything that stood in their path. Three miles beyond the dam, the remains of Hoppy's aircraft were still burning gently, a dull red glow on the ground. Hoppy had been avenged.

Flight Sergeant Leonard Sumpter

BOMB-AIMER, AJ-L

When we reached the area, we came in from the wrong end of it, via the tower, and one of the ack-ack guns had a go at us – I'm not sure if this was a navigational mistake – it might have been. So we came in over this tower, and took several hits.

We circled with our intercom on at about 600 feet in an area to the right of the Möhne Dam, a little way away so that we could see all the action. We heard Gibson say, 'I'm going in now.' We saw him dive down and run over the dam and all the flak and the

Water pours through the breach in the Möhne Dam. 'It was a tremendous sight – the water poured out of the lake, taking everything in front of it. After a few minutes it was just an avalanche of mud and water.'

IWM CH009720.

An RAF reconnaissance photo reveals the devastation, thirteen miles beyond the dam in the Möhne Valley.

guns started firing at him – and he dropped his bomb. We just saw this big spout of water as he came round. So he said, 'Come in number two, you can go in now.' This was Micky Martin.

All the time, we were circling round and saw Gibson and Martin going in and the flak coming towards them – there weren't a lot of guns, but it looked frightening – your stomach began to turn over and the adrenalin was going. You thought, 'Christ, I've got to go down through that!' And there's no turning back. It's not like a soldier who sees a sniper shooting at him and says, 'Blow this, I'm going away.' You've got to go in and that's all there is about it.

While Martin was doing his run, one of the tower's guns were put out of action, so that made it a bit better for the chaps following.

The next one in was Hoppy Hopgood, and whether he was hit or had released his bomb too late – it went over the dam, I think, and it exploded and blew his plane up – and he was killed there.

The next was Squadron Leader 'Dinghy' Young, who made a good hit on it. Number five was Maltby, but when he was making his run, Gibson said, 'It's broken! It's gone!' so Maltby's bomb went through the gap that Young's bomb had made.

We were number 6 and waiting to go in, then Gibson said, 'Right, all that are left, down to the Eder Dam' – which was Les Knight, ourselves and Maudslay's crew.

Flying Officer Edward Johnson

BOMB-AIMER, AJ-N

We watched Gibson attack. He went in first, and we were perhaps a bit shattered at the amount of flak there was – rather more than anyone had anticipated. They were very active and kept it up quite stoically after the bombs had started going off, which I thought was quite heroic, because it must have been quite terrifying to be on the dam wall or in the vicinity when things started going off.

We saw the bombs bouncing quite clearly in the moonlight – at that time our bomb was stationary and the others had been starting theirs up successively as their turn came to bomb, to get it going before their attack. When the Möhne Dam was breached, Gibson gave instructions for those who'd actually dropped their bombs to go back to base. The other three, including ourselves, Henry Maudslay and Dave Shannon, went on to the Eder together with Dinghy Young who was Deputy Controller, and went with Gibson in case anything happened to him on the way to the dam, or when he arrived there.

Flying Officer Harold Hobday

Navigator, AJ-N

We could see fairly well – you could see each plane – but I don't think there were any particular precautions against collision. When we circled, we always did so in an anti-clockwise direction, and we'd been told to orbit at the opposite end of the dam to Gibson's plane. There was great disappointment when the first three didn't break the dam, but the fourth one did – and there were great cheers on board.

Barnes Wallis

Waiting in the Operations Room at Scampton

We had a code by which Guy Gibson, in charge of the raid, would Morse back 'Goner', every time an aeroplane had dropped a bomb, and if you've seen a live bomb explode, you would see an immense column of water that goes up to a height of about a thousand feet. Gibson, after he'd dropped his own bomb, which dropped short and sank short of the dam, would be several miles away and would have seen this tremendous column of water go up – and he'd have Morsed to us, 'Goner'. And if the dam had broken, then the code-word was 'Nigger'. Nigger was Gibson's dog. We had three messages,

'Goner', and obviously the dam wasn't broken, and I could see Sir Arthur Harris and Sir Ralph Cochrane, who was Commander-in-Chief of Number 5 Group, looking suspiciously at me – and I was thinking it wasn't going to work – when the very next call that came through was 'Nigger', and the dam had gone. It must have been a marvellous night for Gibson. He said that after every bomb had exploded, the pilot of the aircraft was to fire a red Very pistol. This immense column of water was mostly spray, and he said it was illuminated blood red by the light of the Very pistol, and you saw a blood red column 1,000 feet high going up. It must have been a very marvellous sight.

Attacking the Eder Dam

Flight Sergeant Leonard Sumpter

BOMB-AIMER, AJ-L

About twenty miles east of the Eder Dam there's another small dam, very much like the Eder only on a smaller scale. I don't know whether Danny made a mistake on giving us the course to it, because it's about forty miles to the Möhne. When we went down and got to this other smaller dam, I said to Dave Shannon, 'That's not it. I'm sure that's not it. It's too small.' So he called up Gibson who came back to us and said, 'Have you found it yet?'

'We've found a dam,' we said, 'but we don't think this is it.'

Then he said, 'No – it's over here' – and he fired a red Very cartridge – and it was about fifteen miles away.

Flight Lieutenant David Shannon

PILOT, AJ-L

We flew east-south-east to the Eder Dam. The flight time was only about ten or fifteen minutes, but it was terribly difficult because by now it was well after midnight, and early summer fog was coming up. It was difficult to navigate because the lake had two arms. I ran down the wrong arm to start with and said, 'I'm not there.' Gibson was right over the dam by then, and he fired a Very light. We saw that and moved over.

The Eder Dam was not protected by any anti-aircraft defences at all. I think the Germans must have thought it was so difficult for anybody to deal with because of its natural surroundings that it was quite secure. It was way down in a very steep valley, and we were flying there above the hills, above the level of water behind the dam, at about 1,000 or 1,500 feet. There was a castle at one end of the lake over which, we judged from our briefing, that we had to drop down immediately over the side of the hill, level out over the water, go over a spit of sand jutting out into the lake and on to the dam wall beyond that.

The Eder Dam was a bugger of a job. For a start, it was much larger than the Möhne Dam. I was first to go; I tried three times to get a 'spot on' approach, but was never satisfied. To get out of the valley after crossing the dam wall we had to put on full throttle and do a steep climbing turn to avoid a vast rock face. My exit with a 9,000-lb mine revolving at 500 rpm was bloody hairy! Then Gibson told us to take a breather and Henry Maudslay went in. On his third run he dropped his mine, and it was the same as with Hopgood: his bomb hit the top of the wall, bounced over, and got the power station below.

Gibson told me to have another run. It seemed pretty satisfactory and we dropped. We made a breach way down under the wall and the rear gunner shouted out that there was a bloody great hole below and the water was pouring out. But the top was still intact. Now only one aircraft remained – Les Knight.

Flying Officer Edward Johnson

Bomb-aimer, AJ-N

The Eder Dam was a bit tricky to find because it was in a very deep valley, surrounded by fairly high territory, and it was a very winding shape. You could fly quite close to it and not see it, so we all had a few difficulties. We found the Eder reservoir OK – that was big enough, but we had problems locating the actual dam, which was in a very restricted area and facing a right-hand turn. Immediately after the dam wall the water-course turned right, which made it awkward to get at.

Flying Officer Harold Hobday

Navigator, AJ-N

We didn't get any flak at the Eder, as the Germans must have relied on the hilly terrain making it impossible for anybody to come down low to attack – and neither dam had any balloons on. There were no fighters up either – but we flew low all the way to upset their radar, which wouldn't work on low-flying aircraft. I'm sure the element of surprise was our salvation.

On our target, the Eder Dam, it was a question of coming over a hill, levelling out very quickly and dropping the bomb – and then going over another hill which was quite high. At the Eder we were the last to bomb because the others had their go. Gibson was there, still commanding the operation. We didn't have much time to orientate ourselves, so we had a couple of dummy runs before we actually dropped it, trying out the beams over the water until we got it right.

As we came in I wasn't tense – I was excited. It was a great thrill if you've been on a bombing squadron as long as we had. I never got frightened – but I sometimes got excited, and on this particular occasion it was marvellous. One learned to keep cool. My job was to make sure the height of the aircraft was right. The wireless operator had his job to do to make sure that the bomb was spun in the right direction at the right speed. The flight engineer had to make sure that the speed of the aircraft was exactly right, and he worked the throttles. The pilot had to steer the aircraft and the bomb-aimer had to use the triangulation device to make sure he released the bomb at exactly the right moment. Meanwhile, the gunners were just waiting for any activity to open fire.

Sergeant Ray Grayston

FLIGHT ENGINEER, AJ-N

To attack we had to come in from 800 feet, drop the machine down to 60 feet, then we had a few seconds to level off and then literally just seconds to release the bomb. Then we had to climb like buggery to get out.

Flying Officer Edward Johnson

BOMB-AIMER, AJ-N

The first one instructed to attack was Dave Shannon, and he made several attempts, perhaps four or five, to try and get down to the level from the high ground over which we were flying – to get levelled out in time to give his bomb-aimer a chance to get squared up and say, 'right', 'left', 'up', 'down' – to get himself in position for dropping. He didn't succeed, so Gibson told him to stop and rest, just circling, and let the second man, Henry Maudslay, have a try. He had two attempts, I think, and didn't succeed.

Then we were instructed to bomb and I think we made perhaps three trial runs before we made our final attack. Les Knight was very good and got us down quickly to a low level, and we had a good run-in. I could clearly see the towers and I was quite happy with my bomb-sight and position – and I released the bomb. I forgot all about the bomb from that moment on, because we were flying directly into this large piece of land, which was only just across the river from the dam. Being in the front of the aircraft, it's quite terrifying to see this looming up at high speed. I was very anxious that we should get pulling the stick back and get over the top which he did, and pushed the throttle right through the emergency gate to get the maximum power – and we skimmed over the top of the hill.

I didn't actually see the dam burst because I was out of sight, being in the front of the aircraft – but it was obvious what had happened by the noise on the intercom from the rear gunner, and everybody else who could see anything was going mad on the intercom, because the centre had fallen out of the dam and the water was absolutely pouring down this narrow river, causing a veritable tidal wave, and we forgot all about safety and going home and we were trying to follow the water down the river to see what happened. It was a terrifying sight. We could see cars being engulfed, then Gibson called up and said, 'Well, it's all right boys, you're having a good time – but we've still got to get back to base. Let's go.'

Sergeant Ray Grayston

FLIGHT ENGINEER, AJ-N

They called back Shannon to do a further run and he was unsuccessful again – he didn't get it lined up straight in the middle, and although he released his mine, it appeared not to do any damage that we could see. We were the sole remaining crew there, and we were lucky really we carried out our first run and we were above the speed permitted, so we aborted the first run, but we learned a lot from it, so I discovered that you had to get the power up to get the speed up to 240, so on the second run I choked the throttles right back to engine idle and let it glide down to the right

Courtesy of The National Archive, ref. AIR20-4367.

The Eder Dam. 'It was still intact for a short while, then as if some huge fist had been jabbed at the wall, a large, almost round black hole appeared, and water gushed as from a large hose.' Flight Sergeant Robert Kellow.

height. There were only a few seconds involved here before you get level and then release – five or seven seconds. As luck would have it, we flattened her out, got the speed right, all the rest doing their job, calling the air speed, looking at the lights and calling high or low, and we were spot on, released the mine and blew the bottom out of the Eder Dam.

I suppose we were lucky – we'd done one dummy run, and got it pretty well right, so we went back round to do it for real. I'd tumbled to the fact that, if I shut my engines right down, she would glide down to 60 feet – which I did. I was responsible for airspeed, so I shut the engines right back and let them idle down to 60 feet, with my fingers crossed that they'd open up when I slammed the throttles forward – and they did. In fact, we were absolutely spot on. We'd hit right in the middle and the bottom came out initially – blew a hole right through it – then the top came away, and that was it.

Flying Officer Harold Hobday

Navigator, AJ-N

There was the thought in mind that the previous bomber had gone in and blown itself up with the bomb – as that might have happened to us – but I think we were confident enough to know that we could hit the right spot. Either their bomb was released too late or else the aircraft wasn't low enough. We were big-headed enough to think we'd be OK. We'd seen their bomb go up – it went with a huge flash – and Gibson called the pilot. There was a very faint reply – very faint indeed – then nothing more was heard. He must have crashed somewhere quite near. Generally our communication between planes was very clear – it was ordinary VHF that we used on the airfield before landing and the like. We only used it when we had to because that would have given the Germans an inkling of what was going on. Shannon went in but his bomb didn't break it. He started trying to give us tips on how to go in. We had to shut him off.

But we made our final run – and the thing broke – and of course everybody was delighted. We watched the water billowing down the ravine from the dam. I could see cars going along, being overtaken by the wall of water. It was fantastic – a sight I shall never forget. We circled to see that everything was going as planned, then we went back to see what had happened to the Möhne Dam.

By this time it was half empty, and we took a look around – and got shot at for our pains. I remember the flak coming from the back of the aircraft, straight past us, but fortunately it missed us – but it was a pretty near thing as you could see it coming, with the bullets whipping past. We decided we'd better make for home, and we were allowed to put the full power of the engines on all the time to make sure we got through as quickly as possible – and we got home without incident.

Sergeant Frederick Sutherland

FRONT GUNNER, AJ-N

Everything's got to be perfect to drop the bomb – the engineer, the bomb-aimer and the navigator calling the height for the pilot – a team effort. And if it's not perfect, you've got to go round again. Like Gibson said, if you don't do it tonight, you're going back tomorrow night to do it again. On the second run we got it right, and as we turned to go up over the hill we could see the water going and a couple of bridges across the river just collapsed. The water was so high it just went over the top of them and they were just gone. Everything was gone. It was a terrific sight and everybody was really happy – we didn't have to come back tomorrow night.

I guess Barnes Wallis was right. One bomb broke it because everything was perfect, and that's the way it was supposed to be.

Flight Sergeant Robert Kellow

WIRELESS OPERATOR, AJ-N

When we passed over the dam wall at the Eder, we had to clear a large hill directly ahead of us. After the mine had dropped, Les pulled the nose up quite steeply in order to clear the hill, and in doing so, I could look back and down at the dam wall. It was still intact for a short while, then as if some huge fist had been jabbed at the wall, a large, almost round black hole appeared and water gushed as from a large hose.

Wing Commander Guy Gibson

In order to make a successful approach, our aircraft would have to dive steeply over a castle, dropping down on to the water from 1,000 feet to 60 feet – level out – let go the mine – then do a steep climbing turn to starboard to avoid a rocky mountain about a mile on the other side of the dam. It was much more inaccessible than the Möhne Valley and called for a much higher degree of skill in flying.

Dave tried and tried again – he tried five times, but each time he was not satisfied and would not allow his bomb-aimer to drop his mine. Then Henry made two attempts. He said he found it very difficult, and gave the other boys some advice on the best way to go about it. Then he called up and told us that he was going in to make his final run. We could see him running in. Suddenly he pulled away – something seemed to be wrong, but he turned quickly, climbed up over the mountain and put his nose right down, literally flinging his machine into the valley. This time he was running straight and true for the middle of the wall. We saw his spotlights together, so he must have been at 60 feet. We saw the red ball of his Very light shooting out behind his tail, and we knew he had dropped his weapon. A split second later we saw someone else; Henry Maudslay had dropped his mine too late. It had hit the top of the parapet and had exploded immediately on impact with a slow, yellow, vivid flame which lit up the whole valley like daylight for just a few seconds. We could see him quite clearly, banking steeply a few feet above it. Perhaps the blast was doing that. It all seemed so sudden and vicious and the flame seemed so very cruel. Someone said, 'He has blown himself up.'

Dave made a good dummy run, and managed to put his mine up against the wall, more or less in the middle. He turned on his landing light as he pulled away, and we saw the spot of light climbing steeply over the mountain as he jerked his great Lancaster almost vertically over the top. Behind me there was that explosion which, by now, we had got used to – but the wall of the Eder Dam did not move.

Meanwhile, Les Knight had been circling very patiently, not saying a word. I told him to get ready, and when the water had calmed down he began his attack. He had some difficulty too, and after a while, Dave began to give him some advice on how to do it. We all joined in on the R/T and there was a continuous backchat going on.

Les dived in to make his final attack. His was the last weapon left in the squadron. If he did not succeed in breaching the Eder now, then it would never be breached – at least not tonight. I saw him run in. I crossed my fingers. But Les was a good pilot and he made as perfect a run as any seen that night. We were flying above him and about 400 yards to the right, and saw his mine hit the water. We saw where it sank. We saw the tremendous earthquake which shook the base of the dam and then, as if a gigantic hand had punched a hole through cardboard, the whole thing collapsed. A great mass of water began running down the valley into Kassel. Les was very excited – he kept his radio transmitter on by mistake for quite some time. His crew's remarks were something to be heard . . . I called them up and told them to go home immediately. I would meet them in the mess afterwards for the biggest party of all time.

Heading for home

Flight Sergeant Leonard Sumpter

BOMB-AIMER, AJ-L

Dinghy Young was shot down on the way home – he was called Dinghy because he'd been shot down two or three times before – always over the sea, and he'd always escaped in his dinghy. He had just crossed the coast when he was hit by flak and came down in the sea.

It was getting quite light on the way back, and when we got near the Dutch coast, Dave said, 'I think we'll go and do a quick whip out in case the coastal batteries are waiting for us.' So we went up to 700 or 800 feet, then he put the nose down and we did maximum speed until we were well clear of the coast, and stayed at low level all the way back to England.

Flying Officer Edward Johnson

BOMB-AIMER, AJ-N

We put everything through the throttle and went back to the Möhne Dam, because that was the approved route for going home. We could see the floods that were forming on the low side of the dam, as the water was still pouring out. We could clearly see that it was emptying – all the sand was showing around the fringes of the lake, even in that short time, and the water was still pouring out in great volume. We could also see that the power station had disappeared below the dam wall. You couldn't see it at all, and there were all sorts of large lumps of masonry lying about down the valley and water as far as you could see.

We set route from there to go home with full throttle, which the engineer, Ray Grayston, said we could use because we'd got petrol and didn't mind wasting it. We had a fairly uneventful journey home – nobody shooting at us. We'd been flying for some time then, it was getting quite late in the night, and we were very relieved to see the North Sea and think that we might be home shortly.

Pilot Officer Les Knight

Pilot, AJ-N

From the AJ-N crew's debriefing report
Routeing excellent. Reports from aircraft ahead re flak found to be very useful. Attack straightforward and as predicted. It was found possible to gain 1,000 feet after dropping the mine. Satisfied the raid was successful.

Sergeant Frederick Sutherland

Front gunner, AJ-N

On the way home, I put a whole bunch in the cab of a moving train in a small town; I thought, well, I'll break the boiler. But the thousand-yard tracer was glancing off – making a beautiful sight, but it didn't do any damage.

Sergeant Harry O'Brien

Rear gunner, AJ-N

We were flying very low during the return journey; at the Dutch coast the terrain rose under us, Les pulled up, over and down. On the sea side of this rise in the terrain, and invisible to Les, was a large cement block many feet high. This block passed under our tail not three feet lower. As the rear gunner, I was the only one to see it.

Flight Lieutenant David Shannon

Pilot, AJ-L

We went back individually, not in formation. I just got down on the deck and opened the throttles and went. The defences were much the same as they had been coming in, with quite a lot of searchlight and flak – but we got through, across the North Sea and back to base. We landed back at 4 am on 17 May. I had been away about six and a half hours.

Flight Lieutenant Harold 'Micky' Martin

Pilot, AJ-P

From the AJ-P crew's debriefing
Very good trip. Numerous searchlights and light flak positions north of the Ruhr against which gunners did wizard work. Rear gunner extinguished two searchlights. Front gunner shot up other flak posts and searchlights. Navigation and map-reading wizard. Formation commander did a great job by diverting the gunfire from target towards himself. Whole crew did their job well.

Wing Commander Guy Gibson

Trevor had fired nearly 12,000 rounds of ammunition in the past two hours, but he said, 'I am almost out of ammo but I have got one or two incendiaries back here. Would you mind if Spam tells me when a village is coming up, so that I can drop one out? It might pay for Hoppy, Henry and Bill.'

'Go ahead,' I replied.

We flew north in the silence of the morning, hugging the ground and wanting to get home. It was quite light now, and we could see things that we could not see on the way in – cattle in the fields, chickens getting airborne as we rushed over them. On the left someone flew over Hamm at 500 feet. He got the chop. No-one knew who it was. Spam said he thought it was a German night-fighter which had been chasing us. I suppose they were all after us. Now that we were being plotted on our retreat to the coast, the enemy fighter controllers would be working overtime.

We were flying home – but we did not know how the other boys had got on. Bill, Hoppy, Henry, Barlow, Byers and Ottley had all got the hammer. They had all gone quickly, except perhaps for Henry. Henry, the born leader. A great loss, but he gave his life for a cause for which men should be proud. Boys like Henry are the cream of our youth. They die bravely and they die young.

And Burpee, the Canadian? His English wife about to have a baby – his father who kept a large store in Ottawa? He was not coming back because they had got him too. They had got him somewhere between Hamm and the target. Burpee, slow of speech and slow of movement, but a good pilot.

I called up Melvin, but he never answered. I was not to know that Melvin had crashed into the sea a few miles in front of me. He had come all the way from California to fight this war, and had survived sixty trips at home and in the Middle East, including a double ditching. Now he had ditched for the last time. Melvin had been responsible for a good deal of the training that made this raid possible. He had endeared himself to the boys, and now he had gone. Of the sixteen aircraft which had crossed the coast to carry out this mission, eight had been shot down, including both Flight Commanders. Only three men escaped to become prisoners-of-war. Only three out of fifty-six, for there is not much chance at 50 feet.

'North Sea ahead, boys,' said Spam. And there it was. Beyond the gap, in the distance, lay the calm and silvery sea – and freedom. It looked beautiful to us then – perhaps the most wonderful thing in the world. Its sudden appearance in the grey dawn came to us like the opening bars of the 'Warsaw Concerto' – hard to grasp, but tangible and clear.

We came to the Western Wall. We whistled over the anti-tank ditches and beach obstacles. We saw the yellow sand dunes slide below us silently, yellow in the pale morning. Then we were over the sea with the rollers breaking on the beaches and the moon casting its long reflection straight in front of us – and there was England.

CHAPTER 8

The Second Wave

The 'Second Wave', whose planned route to their target – the Sorpe Dam – was longer, and their take-off time slightly ahead of Gibson's nine aircraft, began leaving Scampton from 9.30 pm – except for McCarthy, who was delayed by technical problems.

Flight Lieutenant Les Munro

PILOT, AJ-W

I was assigned to the northern force of five aircraft scheduled to attack the Sorpe Dam, which required a different system of approach to the Möhne and Eder. We took a different route in, crossing the North Sea further north than Gibson's wave of nine aircraft. We crossed the Dutch coast at the island of Vlieland, and we came under fire just as I'd crossed the crest of the dunes.

We were flying at 240 mph and I would have been at 60 or 70 feet when we were hit over Vlieland on the port side of the aircraft. The intercom immediately went dead. I felt the thump of the shell. The damage from that shell exploding blew a hole in the side of the aircraft where the squadron code letters were, but didn't cause much damage on the other side and no damage to the rear gunner and his turret because it was probably only a 20 mm. We could see a line of tracer coming up towards us, and it was only a single shell which hit the aircraft.

It was a spur of the moment sequence of events and we carried on down the Zuider Zee, and I asked the wireless op to determine how much damage there was and see if he was capable of joining up the intercom again, and we circled over the Zuider Zee while he was doing that. And he reported back that there was no way of fixing it. I was unable to converse with my crew, which was essential, both as far as navigating to the target and for the bomb-aimer being able to guide me in to the target – so I made the decision to return to base – although of the five aircraft another two were shot down in close proximity and the fourth member, a bloke called Geoff Rice, hit the water with his Lanc and bounced off, but left his bomb and the caliper arms behind and he had to return to base. The fifth member was McCarthy. But it was always no question that I would accept fate – if I was going to buy it, then so be it.

Sergeant Jim Clay

BOMB-AIMER, AJ-W

The sun had set when we reached the enemy coast but there was a little gloomy moonlight. I thought I saw someone to starboard skim the water and send up a plume of spray – it could have been Geoff Rice or Barlow or Byers.

We were over Vlieland when suddenly a flak ship opened up. None of us in the aircraft saw this vessel, although we had, as was customary, been keeping a sharp look-out. We must have been a sitting target to the gunners below – a close target silhouetted against the sky.

A hole was torn in the fuselage amidships, the master compass unit demolished and our intercom completely dead. Les kept on a south-easterly course for a while. Then Frank Appleby passed a short note down to me: 'Intercom U/S – should we go on?' I wrote back, 'We'll be a menace to the rest.' Had it been a high-level operation there would have been time to make up some sort of signals between bomb-aimer, flight engineer and pilot which may have worked, but on a quick-moving, low-level operation like this, and with other aircraft in close proximity, Les could neither give nor receive flying instructions from the navigator nor bombing instructions from the bomb-aimer. A few minutes later we altered course for home – and so ended W for William's effort in respect of this particular raid.

Flight Lieutenant Les Munro

PILOT, AJ-W

I didn't feel emotional about it at the time – it had happened and we had to make the best of it. I was disappointed later on that we didn't get to the target, but by the same token, as it transpired, if I hadn't been hit and returned to base, and if I'd gone on to the target, we may have been one of those shot down. On our return I couldn't communicate with the ground at base, so I had to make a decision to go in and land without permission from the control tower – I had very little alternative.

AJ-T, PILOTED BY FLIGHT LIEUTENANT JOE MCCARTHY

Flight Lieutenant Harry Humphries

ADJUTANT, 617 SQUADRON

McCarthy spluttered, 'My bloody aircraft is u/s. I've got to take the spare. There's no compass deviation card. Where are those lazy, idle, incompetent compass adjusters?' We calmed him down while many willing people searched for the missing compass card. Mac was in a mess. He stood six feet one inch and weighed about 15 stone. The excitement and exertion had really disturbed his equanimity. His shirt was wringing wet and he gulped in great breaths of air. His huge hands were clenching and unclenching spasmodically.

Flight Sergeant Powell came running towards us with the all-important card in his hand – the sound of aircraft engines starting up could be heard.

'Listen now,' said Mac, rushing off with a card in one hand and his parachute in the other. Alas, he was holding the ripcord instead of the proper handle, and as he ran towards the Lancaster yards of parachute silk streamed out from behind him. He didn't stop, but merely flung it to the ground like a discarded dishcloth. Within five minutes of him entering the aircraft his engines were started, and I remember thinking that he may make a mess of the whole thing, taking off in such a state.

Flying Officer Dave Rodger

REAR GUNNER, AJ-T

Joe nearly went without a parachute. In transferring equipment from one aircraft to the other he caught his foot in his chute and it blew open, so he had to send someone running for another.

Sergeant George 'Johnny' Johnson

BOMB-AIMER, AJ-T

In the first place we had a problem with our aeroplane. It misbehaved on the run-up and we had to transfer to the reserve. Now the reserve had only arrived that afternoon. It had been fuelled, it had been bombed up, it had a compass swing – but that was it. It had been armed, so we had to transfer to that – which meant we were twenty minutes late taking off. We went out on the reverse route because that way we felt we could catch up with the others who were so far ahead of us. The inward route was rather more distant than the homeward route, so we went out on the homeward route to cut the time.

Flight Lieutenant Joe McCarthy

PILOT, AJ-T

Very hot reception from the natives when we crossed the coastline at Vlieland. They knew the track we were coming in on, so their guns were pretty well trained when they heard my motors. But, thank God, there were two large sand dunes right on the coast which I sank in between.

Sergeant George 'Johnny' Johnson

BOMB-AIMER, AJ-T

With the modified aircraft, we were toddling along and we saw this goods train just going along very nicely thank you, and Ron Batson, our front gunner, said, 'Can I have a go, Joe?' And Joe said, 'Yeah, OK – right.' So Ron opened up with our little 303s. What we didn't know was that it was an armoured train and, of course, it gave us rather more than we were giving it. So we got out of that. We knew we'd been hit, but we had no idea where. It didn't impede the aircraft anyway, so there was no problem as far as that was concerned. And it wasn't until we got back that we discovered where we'd been hit. In actual fact, a piece of shrapnel had burst the starboard tyre. So that, when we landed we were more than a little bit 'starboard wing down' and had a bit of a jitter going on. But again, Joe managed to control it extremely well. He really was an ace of pilot, there's no doubt about that. He was absolutely first class.

Don and I had an arrangement whereby we both had maps with the track marked on. I would pick out various pinpoints, not necessarily on track, and report back to him, and he would use the information to keep our course. We didn't use the roller map method. Don was of the opinion that should the situation arise where you got badly off track, you would have no map to refer to because the roller kept you to a certain width of track. The ordinary maps took up that much more space, but they were so much better, we felt, for us.

We eventually got to the Sorpe, with a little difficulty in finding it, as it was quite misty. We had to attack across the length of the dam, which meant going down the mountainside on one side and then up and out on the other side quick, before you hit those hills on that side.

What we hadn't been told on the briefing was that, on the line of attack, there was a church steeple on the side of the hill, so we had to avoid that, lifting the wing slightly to get over it. Joe, in his wisdom, decided to use that as a marker and came just over the top of that and straight down. In actual fact, it took us ten runs to get it right. It was a question of getting the aircraft level, getting down to the height, and getting everything right before you needed to pull up and get to the other side. Dave Rodger, the rear gunner, was a bit concerned that every time I said 'dummy run' – which meant 'no good' and he was thinking, 'Oh Christ, what next?' He was getting the full pull of the G, being in the rear turret, and it all going to that end. He said afterwards, 'I was thinking of throwing you out of here.' I learned pretty quickly how to become the most unpopular member of the crew, because I said 'dummy run' nine times. It didn't do the morale of the crew much good, but I knew that if we had to do it, we had to do it properly. So up we went, and round again. On the tenth attack we got it right.

We didn't have the spotlights, but I think we dropped from 30 feet. It had to be a bit of a 'by-guess-and-by-God' attack. The idea was that the lower you got, the forward travel of the bomb was reduced that much more. We lined up the port engine with the dam wall, so that when we dropped the bomb it went down the water side of the dam and rolled down to the prescribed depth before it exploded, allowing us time to get away. At last, bomb gone! And we were up and away. I can imagine the rest saying, 'Thank Christ for that!'

When the mine exploded it was a terrific sight. We were nose up at that stage and turning. The explosion was between our aircraft and the moon and Dave Rodger in the rear turret had a clear view.

I couldn't see, but Dave said the spout of water that came up was absolutely tremendous. In fact, some of it actually hit the rear turret. He said he'd begun to think he was gonna get drowned as well as having been knocked about.' He said, 'God Almighty!' then as we turned, I saw it. By then it was starting to fall back, but it was still a fair amount of water.

As we orbited we saw it had crumbled about two-thirds of the way along the dam wall. We were hopeful – but that was as far as it got.

Although five aircraft were briefed, we were the only ones to get there. Les Munro had been shot up over the islands coming in and the shot had ruined his intercom so that there was no way that they could talk to each other. Another had flown too low over the Zuider Zee and the bomb had been ripped off the aircraft, and the others were lost through enemy action. Barnes Wallis had said that it would need at least six mines to crack it – and then having been cracked, the water would do the rest – because it was such a thick earthenware core, far thicker than either of the others. Unfortunately we didn't have six – we had just one, until later when a second reserve aircraft made it, and again had great difficulty finding the dam. They made an attack but with no positive result.

In theory, it seemed to us all that there was no reason why we shouldn't do it – the dam wasn't defended, we weren't having any

Courtesy of The National Archive, ref. AIR20-4367.

Following the devastating raid on the Möhne Dam, barrage balloons were positioned to deter any future attacks.

real trouble, although it was difficult, we weren't having any real trouble with the flying and there seemed no point in making that journey if we couldn't do the job we had been sent to do.

Heading for home

Sergeant George 'Johnny' Johnson

BOMB-AIMER, AJ-T

As we came home, we flew back over the Möhne – or what had been the Möhne – and it was just like an inland sea. Nothing but water was visible for literally miles. It really was quite a sight.

On the return, the problem occurred because we had to take the reserve aircraft. While there was a compass card in for the deviation with the bomb on, there wasn't one for a swing having been taken before the bomb was loaded, so the deviations were very much different coming back from going out. When the navigator, Don MacLean, noticed that and realised we were using the wrong deviations Joe said, 'Right, we'll go out the same way as we came in.' How we did it I don't know, but we got home. But there were navigational problems and we wandered over the marshalling yards at Hamm. Fortunately we were so low the flak gunners couldn't depress their guns low enough to get at us. As Dave Rodger said, 'Who needs a gun? At this level all they need to do is change points.'

When we came back we landed, and oh dear, the right wing was very low and the hit that had occurred was a piece of shrapnel that went through the undercarriage nacelle and burst the tyre. We didn't find this out until we landed. I also found a second piece lodged in the fuselage underneath the navigator's feet. It could easily have been much more serious, particularly for the navigator.

Flying Officer Dave Rodger

Rear gunner, AJ-T

The starboard tyre had been flattened, unbeknown to us. As we came in to land, though, Joe did a wonderful job. He managed to hold that wheel up by applying aileron until we were just about stopped. We just spun round once. Then when we got out we had a welcome cup of tea. We were relieved to be home.

Sergeant George 'Johnny' Johnson

Bomb-aimer, AJ-T

When we got home I think we were satisfied that we had done what we went to do.

CHAPTER 9

The Third Wave

Unaware of the fate of the first two waves of aircraft, the final group of five Lancasters took off just after midnight on 17 May and set out on the same route as Gibson's wave. These aircraft were to act as a mobile reserve, to be deployed by radio controllers at Scampton, depending on the success of the previous waves against the primary targets.

Flight Sergeant Ken Brown

PILOT, AJ-F

That night, we were perched out on the grass. It was a beautiful night, clear sky, no cloud, waiting for the buses to take us out to the aircraft. John Burpee came over to me before take-off, being a Canadian, a friend, he thrust out his hand and said, 'Goodbye Ken'. I didn't expect he'd come back. You see, some people feel that way. That was not usual – you may think so now, but, we weren't foolish enough to think this was a sure bet, let's face it. If you made this one, you were going to be very lucky.

Flight Sergeant Grant McDonald

Rear gunner, AJ-F

I recall watching the second wave take off independently, followed by three flights of three aircraft each, take off in formation. I had never seen Lancasters take off in formation – that was quite a sight. We still had two hours to wait, being in the reserve wave, and didn't leave until shortly after midnight.

Flight Sergeant Ken Brown

Pilot, AJ-F

We were all frightened. There were so many misgivings. Then we got on board the bus; there were three crews to one bus. The bus stopped to let the first crew off. Then the second crew got off and my tail gunner, when the second crew got off and the bus moved on, was very quiet. Then it stopped for us, so we moved over towards the aircraft. And my tail gunner stood there, where he'd got off the bus. Then he came over to me and said, 'You know those two crews aren't coming back, don't you?' And I said, 'Yeah, I know. Come on, get aboard.' It's happened before on regular bomber ops – don't ask me what it is, but it's just something in their attitude or their lack of chatting – but it was evident that they weren't coming back – and neither of them did. Two hours later they were both dead.

We went out to our aircraft. It was a beautiful night, but no wind. We needed that wind because we were on the short runway and the hedge on the short runway was a thousand feet tall. At least it looked that way when you are taking off.

In most pictures today, they show the Dambusters taking off from runways. That wasn't so. We took off from grass. The bomb that I had on the aircraft was marked 11,960 lb on the side of it.

We got the aircraft in the air and then discovered that we had to use climbing power, 2,650 (RPM) at 9 (boost), to keep the whole thing in the air.

Flight Sergeant Grant McDonald

Rear gunner, AJ-F

The wireless operator got the message that we should proceed to the Sorpe Dam. We encountered some very heavy anti-aircraft fire going through Holland, with the German air force bases there. One of our aircraft was brought down very close to us – then we were caught in the same fire and searchlights. We were hit by flak but we proceeded on and we crossed the Rhine and saw another plane go down, farther away, but we didn't know at this time who they were.

Sergeant Dudley Heal

Navigator, AJ-F

It was a beautiful clear night and we were flying at 150 feet – the idea being that the lower we approached the Dutch coast, the less likely we were to be picked up by radar. We flew over East Anglia and over the North Sea, hit the coast more or less on track, although a little bit off to starboard, then corrected my new course and we flew then over this green countryside on this fine starlit night at roughly 150 feet. I would get up there from time to time and have a look out – it all seemed so strange, everything looked so attractive. Everything was so green over Holland, and again over Germany – and objects stood out – churches, pylons, rivers and canals – so you could tell very easily if you were off course.

Steve Oancia kept me posted of anything he saw that might assist in navigation. Twice, on the way there, once over Holland and once over Germany, we saw a Lancaster go down, a mile or two to port of us, shot down in a ball of flame. The first one was unlucky in that he flew over Gilze-Rijen airfield, which was only a mile or two away from where we were. Then over Germany we flew over the Hamm marshalling yards – which always had been a great target for Bomber Command. But we kept going – we were lucky.

Flight Sergeant Ken Brown

PILOT, AJ-F

Then we came on to the coast, the Dutch coast. Immediately we were in the area of Gilze-Rijen which was a fighter-drome. We all knew that the Luftwaffe nightfighters were there at the time. Pilot Officer Burpee was about a mile and a half off the north coast and they opened up on him and he blew up on the airport. So I knew we had one less.

We went on towards Hamm and I just couldn't help it – there was a train moving along a gentle slope. And I said, 'OK gunners. Here's where you can get your exercise now and your target practice.' So we took on the train as we flew right along side it.

Sergeant Stefan Oancia

BOMB-AIMER, AJ-F

Burpee's Lancaster ahead of us, flew over a German airfield and was hit by ground fire, fuel tanks exploding and the ball of flame rising slowly – stopping then dropping, terminated by a huge ball of flame as it hit the ground, the bomb exploded.

A Dutch eyewitness of Burpee's crash

An aircraft approaches from the west at very low altitude and tries to break through the light flak barrage between Molenschot and Gilze-Rijen. It seems to be caught by searchlights. Then a fire spreading red light becomes visible – the aircraft is on fire and crashes at the airfield among the buildings and hangars. A most terrific explosion follows.

Flight Sergeant Ken Brown

Pilot, AJ-F

We were having trouble, as was everyone else, with high-tension wires. They were our greatest danger at anytime. If the wires in the moonlight were to flutter up just above head level, we knew we'd have to go under them. If they were below head-level, we knew we'd go over them. It was that quick. We lost two aircraft to those wires. They merely slapped into them. Deadly stuff.

As we came along to Hamm they were really waiting for us. We and the other two waves had passed that way. So they poured it down. As a matter of fact, they were firing down at us. They were on a little bit of a lip as we went through the valley. Ottley was on my starboard side at about one o'clock and they hit him. He immediately blew up. His tanks went first and then his bomb.

Sergeant Basil Feneron

Flight engineer, AJ-F

Over on our left we saw some flames and sparks, and a lot of fireworks going off, and Ken said, 'Goddam, that must be Burpee or Barlow' – and it was Burpee.

Flight Sergeant Ken Brown

Pilot, AJ-F

As a matter of fact, seeing where Burpee was probably saved our skins, because had we been over a bit further we'd probably have been the first to be hit.

Sergeant Basil Feneron

FLIGHT ENGINEER, AJ-F

It was a pity – a shame – we felt very upset about it – the loss of life – but in my opinion it happened because they weren't down low enough.

Herbert Scholl

WIRELESS OPERATOR IN JU88

It was 22.00 hours on Sunday, 16 May 1943, when the aerodrome at Gilze-Rijn was crossed for the first time by a low-flying English four-engine aircraft. The nightfighter crews who were on standby were standing in front of their barracks, looking out for the bomber. According to its sound it was getting lower and lower as it stooged around the vicinity. It was assumed that the bomber's crew was searching for a specific objective on the airfield or nearby. The place was completely dark and not even the moon was shining as the bomber approached the airfield a second time. It came from the west just as a searchlight on a tower between the command post and repair hangar directed its beam on to the bomber. The bomber was already very low and the searchlight beam caught it almost horizontally as it approached, presumably blinding the pilot. The bomber dropped even lower into trees through which it tore a great swathe before crashing on to an empty military vehicle garage belonging to the airfield Flak section. It burned immediately. Seconds later there was an ear-splitting explosion caused by a mine. The crash site was about a hundred metres west of the repair hangar, between it and the command post. The shockwave was so strong that it engulfed the nightfighter crews who were standing on the far side, about 600 to 700 metres away. On the following morning one could see that the aircraft was totally destroyed with only the rear turret and tail plane remaining almost undamaged. This had been torn off on impact at its joint with the main fuselage section. The crew was dead. The rear gunner showed no signs of injury outwardly. He was dressed in lace-up shoes with worn-through soles and thin, unpressed uniform trousers. It would have been unfortunate had the bomber flown on for another 100 or 200 metres, because it would have rammed the searchlight tower. Why the flak did not open fire is not known. In any event the successful kill by a searchlight has a certain rarity value.

Flight Sergeant Ken Brown

Pilot, AJ-F

When Burpee crashed the whole valley was just one orange ball. I didn't have too much of an alternative: I don't think there was any bravery connected with it. There was a road off the port side. Everything was trees and this road, I couldn't see because of the fire from his aircraft. So I dove and went along the road. Then much to my consternation that damn road led right into a castle, and I'll never forget that castle door. We had to dip and the left wing went between two turrets as we went through the castle.

We arrived at the Möhne Dam. It had been breached by that time. The gunners were still fairly active. We thought we'd leave them alone and we went over to the Sorpe Dam.

When we got to the Sorpe, the whole valley was filled with fog. The only thing we could see was the village on top – the church spire sticking up through the fog – and that's all we had to go on. We made several runs just to find the dam itself. As a matter of fact, on the third run we almost crashed into it.

Sergeant Dudley Heal

Navigator, AJ-F

As we approached the Möhne Dam we got word to change course and head for the Sorpe Dam. There was no comment but it was a fair assumption that the Möhne and the Eder had been breached.

I gave a new course to the Sorpe, which was very close to the Möhne, and we got there very quickly. We found the Sorpe, the river and the reservoir, and we could see the dam as we flew over the reservoir.

The Sorpe Dam was different from the Möhne and the Eder, which were just concrete dams. With them, the bouncing bomb would arrive at the dam and sink to the prescribed height and explode giving maximum effect. The Sorpe, although it had a concrete construction, was covered on both sides by enormous earthworks so this technique would not work. The only way we could attack the Sorpe was to fly as near the dam as possible, at the prescribed height, and fly along the dam instead of at right angles to it as with the others, hoping that the nearer we could get to the dam and the nearer to the centre of it the better. Judging by the damage that we could see afterwards, we were fairly successful.

We still used the converging beams because we wanted to drop it from the same height – but we didn't use the spinning technique. So at the briefing we must have been told that we were going to go for the Möhne Dam but that we might be diverted to the Sorpe Dam. We knew if we had to go to the Sorpe Dam, we'd have to adopt a different technique.

We turned and came round to do a dummy run. The greatest difficulty we had was in actually finding the dam, as it was completely surrounded by forest and rising ground and mist. Once we got over the forest we couldn't see the dam.

We did about four runs like this until Ken the pilot said to the bomb-aimer, 'Look now Steve, I'm going to go round again and when I give the word, I want you to drop some flares, at intervals,' which he did. When we came round again we could see where the dam was in relation to the flares and adjust our course accordingly. I think we were the only ones who did this, which enabled us to fly over the dam. Even then we had to do another couple of runs – we did about six runs altogether – then we dropped our bomb. I believe it dropped very near the dam – just to the side of it – in the water. We flew on round, and saw the explosion. By this time the dam had already been damaged – we could see that as we flew over. We flew on round, heard the explosion and looked hopefully towards it – but it hadn't gone, even though the damage was greater than it had been before. However, we knew we'd done the best we could.

We were only able to do this because there were no defences at the Sorpe Dam – just the natural obstacles. I think they assumed that anybody who could come and find it and drop a bomb on it was going to be very lucky.

Flight Sergeant Ken Brown

PILOT, AJ-F

It was like driving your car over a rough road and never having your wheels balanced – so everything was bouncing. Your instruments were vibrating – you were vibrating.

Sergeant Steve Oancia

BOMB-AIMER, AJ-F

We had to try about six or eight times before we got the right height, speed and location before we dropped the bomb.

Flight Sergeant Ken Brown

PILOT, AJ-F

So I tried to position myself from the spire. I didn't do too well. I got behind the dam on the first run. When I found myself at ground level, behind the dam, I had to climb up roughly 1,800 feet. It didn't do my nerves any good at all. Because I was on top of the trees, I had to do a flat turn. I couldn't move the wing down to get around. I had to stand on the rudder to get around and then we were down in the valley again.

Well, we did quite a number of runs on the dam, before we were able to clear enough of the fog away with the propellers constantly going through it. However, we were pleased with it and as far as the explosion was concerned, the waterspout went up to about a thousand feet and so did we. I think we ended up about 800.

Sergeant Basil Feneron

Flight engineer, AJ-F

Off she went, and I clearly remember taps full open and throttles on through max boost, and getting out of there – and turning round. Then I saw this waterspout going up – and in those days it was something quite incredible.

Sergeant Steve Oancia

Bomb-aimer, AJ-F

The plume was between the moon and ourselves, and you could clearly see it. It was a great fountain of water coming straight up – which was a reassuring sight – but we looked to see if the dam was still there – and it was.

Flight Sergeant Grant MacDonald

Rear gunner, AJ-F

Although the dam was not breached, there was some damage done. Coming back over Den Helder we got caught in more bad flak and searchlights. The aircraft was hit but the pilot managed to get through it and get us out over the North Sea and back to base.

The Sorpe Dam, being of an earth construction, called for a different bombing approach. 'The only way we could attack the Sorpe was to fly as near the dam as possible – and fly along the dam instead of at right angles to it.' Sergeant Dudley Heal.

Flight Sergeant Ken Brown

Pilot, AJ-F

There was one thing that sort of bugged me. When we went to the Möhne Dam, one of our aircraft, flown by Flight Lieutenant John Hopgood, had been shot down there. And I felt we owed the fellow a visit. So I went back.

All the other aircraft had left. But as soon as we came over the Möhne, they were throwing 20 mm at us. I think there was some that was 37 mm. But I figured that we owed that fellow a visit for John's sake.

Anyhow, we opened up at about 500 yards and carried in over the tower and the rear gunner depressed his guns and we raked the thing as we went through. Well, there was no firing coming from that tower when we left. We figured we'd done him in – however, the fellow got the Iron Cross. So we weren't that successful.

Sergeant Basil Feneron

Flight engineer, AJ-F

I looked down there – because I was further back in the aircraft – and I could see the Möhne. I said, 'Jeeeesus Christ!' And Ken tipped the aircraft so he could see, and Dudley came from behind the curtain to see – and there she was, with the water gushing out. It was incredible.

Heading for home

Sergeant Dudley Heal

Navigator, AJ-F

We turned and headed for home, and sent out the code message which indicated that we had attacked the dam and had not breached.

On our way we flew over the Möhne, which happened to be on our route, so we were able to see that it had been breached, that the water was rushing down the valley. It was a very sobering thought – a very sobering picture. But we felt at least somebody had been successful.

We headed out over Germany and Holland, back over the green land and we were in sight of the North Sea when almost as we hit the coast we were coned with searchlights – at least three – and flak came up from all directions. Suddenly we were in the middle of it. How Ken managed to see in these lights, I don't know, as even in my compartment I had great difficulty seeing anything. But he put the nose down even lower than we already were – which was about 150 feet. As we flew on, we could hear these shells – or bullets – hitting the aircraft, then suddenly we were out over the North Sea. Although, as I stood up and looked out, I could see shells whizzing over the top of us out to sea, but in no time we were clear.

It was all quiet again, and I got up and we went and had a look around and at the body of the aircraft. Just below the roof to one side was riddled with holes, and it was obvious that if we'd been three or four feet higher that would have been the end of it and we'd have been finished. As it was, we were just low enough to avoid the worst of the damage. As we flew over, quietly Ken handed over to Basil, the flight engineer, to take a rest while he came and had a look at it with me.

Flight Sergeant Ken Brown

Pilot, AJ-F

The worst was really yet to come. It was then daylight, or just breaking. We had to go across and up the Zuider Zee. There was no horizon – the mud from the Zuider Zee and the sky were all one. So I started across, strictly on my altimeter with my head below the cockpit top at 50 feet, and I hung on to it. I'd been told by a famous wing commander in the RAF, 'Never, ever pull up. If you're low, never pull up.' So I hoped he was right – because all hell broke loose within a matter of fifteen minutes.

Even though it was light, searchlights caught us from the starboard side and straight on. There was a lot of light flak immediately in front of us. The cannon shells started to go through the canopy, the side of the aircraft was pretty well blown out, and there was only one thing to do. That was go lower, so I put her down ten feet. We came across and actually their gun positions were on the sea wall. So they were firing slightly down at us and I guess they couldn't believe we were lower than what they could fire. So in this turmoil with the front gunner blazing away at them, I just got a glance, for a moment, and I could see the gunners either falling off because they were hit from our guns or rather they were jumping off to save their skin.

I pulled up over the top of them and we all gave a great sigh of relief. We figured we had it made at this stage of the game.

I called each of the crew members and I was really surprised to find that no one had been hit. There was a great deal of damage.

My wireless operator said, 'Hey Skip. Come on back and crawl in and out of the holes.' I did go back. I wondered how badly and what damage had been done to our landing gear etc. But by that time we were in broad daylight. I'm sure that the Germans figured that we were a Kamikaze crew to do what we did.

Flight Sergeant William Townsend's crew

Pilot Officer Lance Howard

Navigator, AJ-O

I had visions of the bumpy take-off causing the lights under the fuselage to be shaken loose so that instead of being 60 feet above ground, we would finish up 60 feet underneath it.

Flight Sergeant George Chalmers

WIRELESS OPERATOR, AJ-O

Once we'd been briefed we went down to the aircraft and got on board and we took off in the third wave at 12.10 am. There were two waves ahead of us who took off earlier.

At 100 feet, incidents came and went almost before the mind could appreciate them – flat meadows sped past as we thundered over Holland and Germany. We attracted some attention when we tried to cross a canal. We did a quick about-turn. Looking out I could see treetops and realised we were circling a forest. Some debate took place about the best method of approach, and the considered approach was, 'Head down, keep low and go on through.' It worked and we were safely through with ack-ack all round us.

Sergeant Douglas Webb

FRONT GUNNER, AJ-O

The fact is that if I had not borrowed an extra 1,000 rounds for each gun and rearmed while flying, I would have had no ammunition for the return trip. The other fact is that if it had not been for the absolutely superb flying that Bill put in, simply going lower and lower, we would not have survived. It is as simple as that. I still remember very vividly some of the power cables and pylons. Bill was the best pilot I ever flew with. We fought our way in and we fought our way out.

Pilot Officer Lance Howard

NAVIGATOR, AJ-O

Townsend threw that heavily laden Lancaster around like a Tiger Moth and we flew out of it.

Ahead and to starboard searchlights broke out and an aircraft was coned at something over a hundred feet. More searchlights and lots of flak and a terrific explosion in the sky. It was one of ours – probably a little too close to Hamm, a heavily defended rail centre, and too high.

Flight Sergeant George Chalmers

WIRELESS OPERATOR, AJ-O

Every aircraft had to reply to group messages – which were transmitted every half hour – simply by sending back their aircraft letter. So as group headquarters sent a message out or simply just made a transmission we replied in order of aircraft: letter A would be first and B second and so on – I was in O for orange. I found it a little bit disconcerting after a while to realise, as I listened, that certain letters were not appearing. I had no idea whether they had been shot down or didn't hear the transmission or didn't make a transmission themselves. But as we got into the heart of Germany I saw large explosions in various places – which were obviously aircraft going down – being shot down or crashing for one reason or another. I didn't know who they were, but I believe from my skipper that there was a Canadian who should have been on our starboard side who crashed.

Our dam was the Ennepe and as we approached it we suddenly saw all this water flowing down. Then at almost the same time, we got the message over the radio that the Eder Dam had been breached. I got the transmission that 'Dinghy' had been transmitted, so we knew that a dam had been breached and there was quite a lot of excitement in the aircraft.

Now we realised the water was coming from the Möhne, and it was quite remarkable and spectacular to see all this water flooding down through the valley. All you could see were treetops and tops of houses. I thought it was just a miraculous sight to see all this water. I don't think we ever gave a thought to the people who would be drowned – at least I never did. There was a job to do and we just did it. We spent a little time looking at that, then carried on to our own target.

We slipped over the top into this big reservoir at the Ennepe – it was very misty. I could see a bunch of trees on the hillside, and we were heading towards that. Next think I could see us going down and there was a conversation going on between Bill, the navigator and the front gunner and bomb-aimer. The bomb-aimer was having difficulty lining up his instrument with the towers. Anyway, this was just the first run in and there was no opposition at all. Climbed up, came round and started to come in to do it again and he was still having difficulty lining up his instrument, and I think they might have some concerns if this was the right one – but it was the right target as far as we were concerned. If it was the wrong target we were told the wrong one.

The weapon started spinning, and the whole aircraft was shaking – the vibration was tremendous – it was a bit frightening at the time. The whole aircraft was shuddering. Lance went down and said, 'Is it turning?'

'Can't you feel it?' I said.

'Yes, but is it going fast enough?' he said.

'Well, there's a red line on the thing,' I said. 'Look.'

'OK,' he said, so I left it.

Pilot Officer Lance Howard

NAVIGATOR, AJ-O

With the aircraft shaking horribly with six tons of bomb revolving underneath, I had to lean in the blister on the starboard side and guide Bill to the correct height with the two lights.

Flight Sergeant George Chalmers

Wireless operator, AJ-O

As we approached our target, we did quite a lot of steaming around to find out just which dam was which – there was more than one in that area. Finally we decided we had got the one we had as our target.

There were a lot of dams around, so it was very difficult to identify our one – but as far as our skipper was concerned we found the one we were briefed to attack so whether it was the right one or not it was the one we were briefed to do. The navigator was quite convinced it was the right one so I took his word for it. It was just like a big lake with trees and land around it, and quite hilly in parts. You had to get over the hill to go down on to it. Initially there was no flak – we'd had all the flak prior to it. But at one stage we were crossing a canal to reach the dam area, and all hell let loose, so we quickly did a complete turn about and flew around a forest and approached from a different angle. I think we were lucky to get through. That was the last bit of flak we actually suffered.

Our wireless equipment worked very well – although I believe from our signaller's leader that I did miss a couple of transmissions – possibly during the run-in or when I was doing something else. My log book would have recorded that I didn't receive them.

We dropped the mine, but it hit the water and apparently bounced only twice, then it must have rolled along and found itself at the bottom of the dam and exploded.

I was in the astrodome during the actual run-in. One of my functions was to start the hydraulic motor to turn the mine and I was sitting on the floor of the aircraft watching the rev-count. When it reached the appropriate revs, then it was ready for dropping. Once it dropped, I got up to have a look – as soon as the plane passed over the wall we started climbing – and that's when I saw the wall of water go up in the air. Quite spectacular.

We were flying over the rooftops of Holland. Dougie Webb says, 'Can we have a pot shot at some of these greenhouses and stop the Germans from getting their tomatoes?' and Bill says, 'Yes, go on, I'll give you a hand.' As we got into the Zuider Zee, the Germans had an ack-ack gun trained on us. You could see the shells bouncing off the water as we went through the gap and into the North Sea.

We were quite surprised that there didn't seem to be any fighter opposition, but after the raid we learned that our own fighter boys had been attacking the fighter airfields that might have given us trouble.

CHAPTER 10

Below the Broken Dams

When only 11 out of the 19 Lancasters which had set out on the dams raid returned, 617 Squadron felt the loss of the 56 men who didn't return keenly. All were presumed dead, although three survived as prisoners-of-war. But in the valleys below the breached Möhne and Eder dams, the villagers suffered much greater losses as the waters gushed with enormous power through their homes and farms, carrying all before them in a torrent of bodies and debris. When the Möhne Dam was destroyed, 1,650 people lost their lives, including 749 foreign workers in a labour camp, and a further 51 people died in the Eder floods.

Johannes Doerwald

I belonged to Home Flak. Our position, equipped with three 2-cm anti-aircraft guns, was situated near the industrial harbour in front of the River Rhine lock. Earlier in the evening, aircraft had flown in the direction of the Möhne Dam and returned without their bombload. At about three o'clock in the morning, a single Lancaster came in, flying very low. The aircraft was fired upon and a fire broke out. It came down near Klein Netterden. The crash took place in a meadow east of the Industrial Brickworks along the Osterhold Road. I was a gun layer and I received a medal for shooting down the plane. Afterwards we heard that officers of high rank were on board the plane. The whole crew, about seven or eight men, were killed.

Herr Feldmann

The aircraft returned from the Ruhr in the early morning. It was fired upon first by the anti-aircraft battery in the keep. Along the Nierenbergerstrasse and near the harbour there were 2-cm anti-aircraft positions of the Home Flak. Each battery had three guns. The plane turned away and the rear gunner fired at the battery near the lock. Then all twelve guns were shooting at the plane and the engines caught fire and it crashed and exploded. People said there were seven crew members on board. Our battery was situated on the harbour breakwater. The aircraft was flying at such low level that our gunfire cut the tops off the poplar trees standing near the harbour.

Unteroffizier Karl Schutte

COMMANDER OF GUN CREW ON NORTH TOWER OF THE MÖHNE DAM

We stood with our guns at the Möhne Dam in total peace. Only after the removal of the remaining batteries and the barrage balloons, which were more urgently needed elsewhere in the Reich and at the front, did our battery commander, Lieutenant Widmann, remind us, 'Children, children, you wouldn't believe what a responsibility you have – how much depends on the Möhne Dam. The whole Ruhr region is supplied through it.' Widmann simply couldn't believe that such puny protection should be given to such an important target.

On the night of the 16th, it was very quiet – apart from the solid, even footsteps of the watchmen in the tower above us, there was no noise. Then the phone rang from the flak post. At Schwansbell Castle at Lünn they'd sounded the alarm. So the lookouts alerted the gun crews and in seconds the defence was ready to fire. We waited, a bit bored, and watched the shimmering reflection of the moon on the surface of the lake. Suddenly, from the north, around Soest, the noise of engines grew nearer and we believed it was from a passing aircraft, using the lake as a navigation point.

Since the distant engine was still nowhere to be seen, we sent up some random defensive fire in a mix of two explosive shells to one anti-tank shell, as it was apparent that the aircraft were very heavily armoured and could never be shot down by ordinary shells alone. We'd scarcely delivered the first salvo when the attacking aircraft started firing. Now, at a distance of some 1,800 metres, we could just make out the aircraft as black shadows. Already more aircraft were circling around the lake and our defences were firing in all directions.

Suddenly we saw one aircraft send out two powerful beams of light on the lake as it hurtled low towards the middle of the dam wall. By doing this they gave us a good marker as to their position and we no longer needed to aim towards the sound or the shadow. Our batteries were better fixed on their target. Suddenly the speeding black shape was thundering like a four-engined monster between the two towers and over the wall at a height of about 20 metres, spitting fire and almost ramming the defence post with its tail.

Moments later there was a horrendous explosion and a 300-metre column of water thundered into the air from the lake. Waves as high as a house battered the crest of the dam. Now we knew that it was us they were after. Still, we had no time to wonder at this amazing spectacle – we kept the aircraft under fire as they roared past.

Change target – more incoming! It came roaring at us out of the moonlight like a monster, as if it wanted to ram us on the tower – but we didn't think about the danger. While my crew commander directed the fire to their height, I stood behind him, giving corrections and directing the action. We fired for all we were worth. The attackers lashed in their fire, almost in our faces. 'Harder! Harder!' I yelled at the gunner who was directing the fire, 'You're getting to them!' And then I clearly saw the streams of tracer drumming into the aircraft, and moments later a flame billowed out. 'It's burning! It's burning!' I yelled to my men, as the aircraft rushed past us, trailing a sheet of flame. Then once again there was a massive explosion. The dust almost took our breath away. Stones flew about us and a huge impact threw me to the floor. As I got up, I saw that nothing else stood of the power station. To the north-west there was a great flash. The ensuing explosion meant the aircraft had crashed.

Still, we had no time to take in the wider picture around us. We swapped the overheated gun barrels and oiled up our weapons. Apart from the man directing fire and the munitions gunner, everyone was on rearming, as the previously armed-up gun was almost shot away. Everybody did what they could. Then a report came from the south tower. The gun post was finished – it had been shot away from the base. Now there were only two small guns to engage the aircraft on the water-side, as the guns down in the valley could only fire behind the aircraft as they turned away.

The sound of engines swelled again after the smoke had dispersed. The swooping aircraft once again turned on their lamps, enabling us to focus on them. We opened fire – which the aircraft replied to immediately. Now a third aircraft came in a direct attack. We changed target on the incoming four-engined monster – and once again the cannons spat out their fire against the attacker, and again the aircraft took countless hits. But what hope had such light 2 cm guns against such a massive aircraft? Only a chance bullseye could do any good. The defences on the banks fired after us and their tracer fire came towards our tower as if on a string of pearls. One could easily have thrown in the towel in face of these overgrown glowworms – but duty had to come before our personal safety.

Then a big explosion and an amazing column of water. Once more the Möhne Lake shook, and once again, great waves beat against the crown of the dam. We didn't know if the wall still stood – we just fired and fired. Once the mirror of the lake had settled again, a fourth aircraft came in on the attack. Our tracer hit it in the fuselage and this together with the attacker's own fire, made strange geometric shapes which were mirrored in the surface of the lake. One couldn't take in any more, as the roar of the engines combined with the sound of our own fire and the explosions, which doubled as they echoed round the valley and appeared to come from all sides. As we could see only the aircraft, we didn't observe when and how the bomb was released. Luckily the wall still stood after the fourth attack. But we no longer knew which aircraft we should engage first, as the four-engined monsters were flying in together to attack. Could we possibly prevent the enemy from releasing their bomb? Already we'd been fired on by some aircraft at the sides of the valley. We replied to their fire – then suddenly our weapons stopped. Our guns fired no longer – the end looked close. We tried desperately to clear the jammed guns, but all to no avail. We just despaired.

So we stood high on the tower – in front of us was the lake and behind us the valley – but we couldn't fight back any more. We now had attacks from all directions. We waited for the end. Then the fifth bomber came in on the attack and flew in over the middle of the dam with furious speed – it was child's play for him now. The aircraft was close enough to touch – I believe to this day that I could see the outline of the pilot in his cockpit. But our guns were silent. And now we did what we had practised so often, attacking the aircraft with carbines out of pure self-defence. It seemed almost laughable – but we fired and it took our minds off the danger we were in.

Once again the Möhne Lake quivered and a giant wave built up under the walls. As visibility got a little better I saw signs of a breach on the wall and I shouted to the lieutenant, 'The wall's a goner!'

'No – that can't be!' he replied.

But I saw that the water was streaming towards the centre and the hole in the wall was getting bigger. I thought, 'If another attack comes in now, we'll all end up down there in the maelstrom.'

The water gushed unstoppably through the breached wall into the valley, and the air was so full of spray that practically all visibility was gone. The swooping aircraft turned away. They had done their job.

Max Schulze-Sölde

Artist, resident on a hill above the Möhne

I watched as a bomber flew deep over the dam, and a little later an enormous mushroom cloud of smoke rose up just in front of the wall. A few seconds after, the sound of an enormous explosion reached us.

The air pressure was so powerful that I was thrown right back into the house through the open door. After a few minutes' calm another aircraft appeared over the wall. The flak guns engaged it and suddenly a flame shot out from it and the stricken bomber streaked close past us like a giant torch. Then the power station went up! I saw the burning bomber fly over the mist and disappear behind a hill where it crashed in a massive explosion. Once again they flew at the dam, and again great spouts of water shot up. Then a great roar rang up through the valley towards me, the small lake in front of the dam grew wider and wider and the landscape suddenly seemed completely altered. Powerful waves shone out silver in the moonlight – the wall was broken! I rushed down to the village of Günne then back to the dam wall. It was an unforgettable sight – deep down the waters of the lake moved like polished metal and resolved into a chaotic, swirling witch's cauldron.

Wilhelm Strotkamp

CAPTAIN OF A PLEASURE BOAT ON THE MÖHNE LAKE

I was recruited as an assistant policeman and put on watch over the Möhne Dam. I was on watch on 16 May, a very bright, warm, moonlit night. Around midnight I heard the bellowing of a siren in the distance and I immediately knocked on the window of the power station and shouted to the switch attendant, Clemens Köhler.

He immediately got on the phone. Then, at first far away, I heard the noise of aircraft engines. The planes came in with a steady power – and so low that one could see them quite clearly. I quickly ran back to the power station to warn Köhler. Whether he heard my shouts I don't know. The first aircraft circled around and bombs were falling into the water. Instead of running up the slope out of the valley, I sought shelter in a tunnel under the wall – which could have been my downfall.

The flak flew like crazy and more bombs fell in the water, which bellowed out and foamed over the wall. I thought the aircraft were trying to destroy the torpedo nets so as to be able to strike at the wall. I became aware of the danger and as I went to get out of the tunnel, an aircraft from the side of the valley targeted the ack-ack post and I was forced to stay put.

After the attack was over, I raced over the great lawn area by the power station and up the side of the valley. In that moment, I heard an aircraft come in again and fly between the towers – and shortly after that there was a resounding explosion, combined with a blinding flash.

I was thrown to the ground by the enormous blast and saw that the power station had taken a direct hit. With an unholy racket, great chunks of broken masonry flew around the place. I couldn't explain the dazzling light.

I quickly crouched down behind a thick tree trunk and after a while I heard the dull sound of engines again. This time, however, there was a completely different-sounding explosion. Suddenly the earth I stood on shook and the whole wall trembled as if a thresher was going through it.

It was like an earthquake and the wall shook between the towers and trembled from the lake to the valley side. Then I saw the water gushing through the splits in the wall structure – that was just for seconds, moments – and then the whole lot broke through like a barn door opening.

Now all I could hear was an incredible roaring and crashing and in the next seconds I found there was water all around me, and I couldn't get up the mountain fast enough as the water surged on. Out of breath, I reached the sentry post – from which all the windows had been blown out.

The watchman grabbed for the phone, which had that moment started ringing. The district administration in Soest wanted to know what was up. We reported the break in the wall and said, 'We can't do anything here. You've got to warn them down in the valley.' With that we passed on the responsibility.

When I got back to the dam wall there was nothing to be seen – just mist and fog raised by the enormous mass of water gushing into the valley.

Herr Tücking

Builder, living in Marbeck

It was around 11.30 pm when I was shot from my bed by the noise of a low-flying aircraft which nearly took the roof off the house. I ran into the garden and saw that another aircraft had flown into a pylon and had gone up in flames. The top of the pylon had been thrown into my neighbour Thesig's yard. There was a crash and everything was lit up. At that moment, all the guns on the burning aircraft went off. The aircraft slithered right across the neighbour's yard and exploded in the meadow. Ammunition was going off in all directions and our mare and foal crashed through the fence in their panic. Then I watched from my hiding place behind a wall as a fiery ball burst from the wrecked aircraft and rolled some 150 metres away (probably a Wallis bomb, saturated with fuel and on fire!).

The next moment there was a fierce explosion as if the whole house were going to collapse. It was only after about half an hour that we could approach the scene of the wreck, where ammunition was still exploding. As we ventured closer, we saw a man in a crouched position, leaning on his hands close to the wreck. He was stiff and completely charred. A gruesome sight. Then we went further on to where the bomb had exploded. On the way there was nothing to see, but on the edge of the crater, 150 metres away from the wreck, we found three or four young airmen, all dead in their thin uniforms, without any outward signs of injury. The pit left by the bomb was 12 metres across and unbelievably deep. By a miracle, a memorial with a figure of Saint Joseph which stood in the immediate vicinity was not damaged, but within a radius of 3 kilometres the roofs were off and the doors and all the windows shattered.

Herr Schulte

Reserve policeman in Ampen

I was woken by the noise so I set out towards Ostönnen. En route I met a Hitler Youth lad, who had arrested an airman who had jumped out of a stricken aircraft.

I took the airman with me to my duty post in Ampen and questioned him as well as I could. He explained to me that it was his twenty-first flight and he was an Australian pilot officer, and he would have got 20,000 Marks if he'd got back from this mission. When I asked him what sort of aircraft he was in, he refused to give any information. He offered me cigarettes from a large case and we talked some more.

After a while I flagged down an army vehicle on Route 1 and I was driven with the captured airman to the airfield at Werl, where there were already four other airmen sitting awaiting interrogation. I couldn't say if they'd all been taken on that same night, but you could tell by their clothing that they hadn't been there for long.

I recorded in my notebook:

'A police search was undertaken for any other men who had bailed out. It was established that the airman had bailed out of an English aircraft which had crashed to the left of the Ostönnen-Volbringen road. Of the seven-man crew only two had survived the jump. The captured airmen were called Burcher and Fraser.'

In the morning the regional head of police from Soest came to me and ordered me on to blockade duty at the Möhne. The surrounding region had been declared a disaster area and it was strictly forbidden to enter it. A colleague and I controlled the area between Niederensee and Günne. We had trouble holding back the many curious sightseers, and above all, preventing them from taking unauthorised photos. A lot of cameras were confiscated.

From the height of Günne I could see the wedge-shaped breach in the wall of the dam, with a huge expanse of scree before it. In the top boughs of a tree hung a piano. We had a lot to do in order to stall the people who were now thronging through the fields in the direction of the dam wall.

Eberhard Viegener

PAINTER LIVING IN SOEST

I was visiting friends when I was surprised by the air-raid siren, and we went outside on to high ground with a view over the Möhne valley. Ten minutes after the appearance of the aircraft I heard the first explosion. Then it was quiet for a few minutes, until we heard the roaring of water. A monster wave about ten metres high was rolling in to the east, then flattening off somewhat towards the south. Instantly churches and estate buildings stood up to the roof trusses in water. You could still see the tops of the steeples reflected in the water, but soon they tilted over to one side and disappeared.

Uprooted trees were carried along on the waves. It was a terrible feeling to know the water was so close below my feet. Railway tracks from the branch line had been slung around the trees like creepers. Trees, uprooted and stripped of bark looked like ghosts.

Elfriede Baader

HOUSEMAID AT HIMMELPFORTEN FARM

Shortly before one o'clock Frau Schewen came out of the farmhouse cellar. Herr Kersting, the tenant, had yelled to us, 'Save yourselves! The water's coming!' We ran east along the road from the farm up to the woods. Between the buildings we encountered the first lashing of the house-high waves, which were accompanied by clouds of water vapour. Here, Frau Schewen was ripped from my grasp, and a woman who was visiting with her three-year-old child, was carried away by the waves. On the road I watched as waves engulfed the pastor's house and the church. There was an indescribable crash and a bursting of walls and beams. I saw only water vapour and rushed into the wood, where I was reunited with the Kersting family.

Elizabeth Hennecke

YOUNG WOMAN LIVING IN NIEDERENSE

On the evening before that night of horror, we youngsters went by bike to the cinema in Neheim. We'd hardly got home when the alarm went off. We took the boxes containing our papers and valuables down into the cellar. It was nothing unusual for us, as the enemy aircraft flew fairly low and were always circling over the Stetsberg. We went outside to have a look around, to see if they were dropping fire bombs. Suddenly there was a fearsome explosion. We fled back to the wall of the house – but the light inside had gone out. In fact, they'd hit the power station. We ran down into the cellar and lit some candles.

It was quiet for a little while, but then we heard an ear-splitting roaring, cracking and crashing. We ran up again, and found the water already knee deep in the yard. We screamed, 'The wall's been hit!' and ran back into the cellar to fetch the children and old people. Then the cellar just caved in. We raced upstairs, but the water was almost as quick as we were. We ran up another flight of stairs to the ground floor. There, by good fortune, we had a ladder which the chimney sweep always used. We used this to climb up to the smoke trap, tore away the rafters and pantiles and climbed up to the attic. Below us was the seething water, full of wood and scree – the children cried and screamed – a few were praying. I could do neither – my throat was completely dry. So we knelt under the roof until four in the morning. We signalled with our pocket torches and the people who lived up the mountain saw that we were still alive. Around nine or ten in the morning, they came through the water and metre-deep sludge with ladders and got us down. They gave us some dry clothes and a warm drink.

Werner Kittler

DIRECTOR OF RESIDENTIAL AND INFIRMARY COMMITTEE, UKRAINIAN WOMEN'S LABOUR CAMP

Around midnight an air-raid warning went off. There was a full moon – you could read the paper by it. Then low-flying aircraft came and shot up the barracks. So I sounded an all-out alarm and gave instructions through my male and female interpreters that they should position their work comrades in the shelter of the barracks. My faithful sheepdog, Olly, was amazing, and simply couldn't be pacified. He wanted to drag me away from the camp. Then I suddenly heard a fearsome roaring. At first I got the impression that it was the light railway which went from that side of Neheim to the Möhne Dam. All of a sudden I saw, some distance away, a gigantic wall of water approaching. I thought, 'They've broken through the dam wall!' So again I gave instructions through my interpreters that they should tell their people to run up through the Möhne River below up the nearby Wiedenberg. But they lost their heads and a bunch of them started running at me in a panic. I couldn't hold them back any more. I gave my interpreter a pistol and I took my pistol and told them that if anyone charged at me again, I'd just open fire. By doing that I was able to save hundreds of lives.

Then amid the screaming, I suddenly saw a boat come whooshing past me towards the barracks like an express train – but the barracks had gone. The whole performance had taken just five to ten minutes. I managed to get myself on to a five-ton steamroller which was standing in front of the camp, and to chain myself on firmly, with my sheepdog secure on his lead. Then a beam banged into my arm and I couldn't hold on to Olly any longer, and my faithful dog was whisked away from me by the floodwaters. I lost consciousness and later found myself on a piece of timber from the barracks, drifting in a deserted corner of a factory, surrounded by drowned bodies. How I survived all that is still a mystery to me and was truly a miracle. For me it was a small consolation to have at least been able to save the lives of a few hundred Ukrainian women.

Frau Gustel Schulte

SURVIVOR

The waves just swept over anyone on foot, walking or running in front of them, but we'd seen the waves, and managed to escape.

But it was terrible – the cries and screams. The other tragic thing was that at the foot of the Wiedenberg – that's the big hill up near the Möhne river – there were labour camps where the Ukrainian women were held. Most were locked into their barracks and couldn't get out. They were trapped inside the barracks as they were swept down until they came to the concrete bridge, where they were smashed to pieces. The terrible screams of the women still trapped inside still rings in my ears. The screams were terrible – just terrible.

Rector Mahnschulte

NEHEIM SCHOOL RECTOR AND GUARDIAN OF THE LOCAL MEMORIAL

Our house stood on a steep mountain road, from which we overlooked most of the town. The view over the Möhne Dam was obscured by a mountain ridge. On that dreadful night the sirens wailed around midnight. No more than a quarter of an hour later I was woken by the noise of aircraft, so I knew something was up.

I took my wife and our four children with me into the air-raid shelter and ran to the window. The sky was dotted with red star shells and the noise was getting louder. A few minutes later I saw a burst of flame in the sky. A bomber was reeling towards the earth. On impact there was a fearsome explosion. After that there was a different noise. Soon the air was filled with it. People were running on to the street and standing there in shocked groups, staring into the valley. Suddenly we could hear cries and screams over the thundering noise – the screams were coming from the labour camp. I called to my wife, 'Quick, get the children out of the cellar. The dam's breached!' We ran into each other in the dark as we groped about for money and valuables. As we ran on to the street, I called out as loud as I could, 'The dam's breached – run for your lives!' My neighbours knew it too, as hundreds of terrified women, children and men ran up the mountain from the lower end of the road.

The noise of the water crashing in was ear-splitting. As I turned to my wife, I saw her and the children hurrying up the mountain among the crowds. Instead of following them, I took a route that I knew.

Then I saw the flood – an enormous, foaming wall, hurtling towards the town at the speed of an express train. In the narrow valley this tidal wave must have reached a height of ten metres.

In the greyish black mass, telegraph poles, bits of masonry and furniture were tossed around. It roared along, ripping up trees and houses. Electricity pylons were picked up and submerged into the water with blinding flashes. Then it was past, leaving me behind, stranded. Totally exhausted and sick with fear, my head throbbed from the noise. I waited until the water had receded enough for me to set out through the debris and ooze towards my house. I passed groups of people who, blind with terror, were calling out the names of friends and relatives.

My cellar was completely flooded. If we'd stayed there, we would have shared the fate of so many of our neighbours. Vainly I tried to console a little girl who'd escaped that fate because she'd left the cellar because of a headache. Eleven members of her family had perished. Soon I established that my wife and children were safe. I ran back up the mountain with my wife and gazed down into the valley. I saw small groups of people stranded on their roofs, the bodies of those who had drowned, among them many foreign labourers. A quarter of the town was just one huge lake.

A local Nazi commander was strutting around nearby. I heard him say, 'To be sure, the Brits have achieved a brilliant piece of work. For doing this a German would receive the highest accolade.' I couldn't believe it. Someone was breaking Hitler's sternest command, in that he admired the enemy. If he could allow himself to do that, I thought, so could I. So that morning I took photos of the devastation – an offence which could have had me put in prison. Soldiers and aid workers were now marching in.

About 8.00 am a junior officer came to my door with eight men. They'd got orders to take me to help in the operation to clean up cultural landmarks – that was my task. Religious statues and other valuable artefacts had been strewn over kilometres by the water. The scene which the receding waters had left behind was indescribable. Bodies everywhere, buildings just ruins, thick sludge all over the streets. Thousands of people worked day and night to provide drinking water and return the electricity supply. We also learned that the Eder Dam had been breached at the same time, so we in Neheim were not the only ones overtaken by this calamity.

Herr Mette

RESIDENT OF NEHEIM-HÜSTEN

I lived with my wife and children in a house at the sports ground. I was woken by a crashing and roaring all around. Was the house shaking – or was I imagining it? I opened the window. The house shuddered and I was thrown out through the window and fell straight into the water. I was drawn down into the depths by the swirling water, then it spat me out again. I grabbed a tree trunk, which was ripped away from me just seconds later, then eventually I caught hold of the bough of a very tall tree. I stayed sitting there all night. Next day the rescue party got me down from the tree and asked me how I'd stuck it out. My answer: 'If I hadn't I'd be dead too.' I'd learned later to my horror that my wife and children had drowned and our house had been washed away.

Josef Rösen

MACHINE OPERATOR IN NEHEIM

The sky in the northeast was lit up in a flash by a fiery glow. My first thought, 'It's the dam!' I ran into the house as fast as I could and woke my sister, and asked her to get some drinking water and

After the flood waters had subsided the full extent of the damage in the Möhne Valley was revealed. A house in Neheim has been ripped open by the avalanche of water.

candles, as water and light were bound to fail. I heard an extraordinary roaring. For me there was no longer any doubt that it was the water from the dam coming down the valley. Over the roaring I could hear people screaming and animals bellowing. I went on my bike to the town. The water was already a metre deep around the Neheim church, so I couldn't go any further. My aim was to get to the abattoir, where I wanted to help my mates who were on duty. The night was so bright that I could see a pig running frantically up and down on the roof of one of the sheds. I turned round and tried to get through to Friedrichstrasse. I had to push my way through joists and boards and other floating debris in order to make any progress. Suddenly I stumbled on the unclothed bodies of four young women in the scree. I stowed them behind the nearest door and cleared my way further along the road to the flooded railway embankment. In the meantime the temperature dropped to about three degrees. Mist was gathering over the water.

Eventually I got to the abattoir. I found more bodies among the roots of bushes and trees in the millrace, and I covered them over. Then suddenly I heard voices coming from the mist. I looked up and on the fork of a branch I could make out human figures – also completely naked. They were young women – foreign workers from the camp beyond the abattoir. The rushing water had ripped the clothes right off their backs. I was able to get four or five Ukrainian women down out of the tree alive and into safety.

I toiled on along the railway embankment to the house, up to my waist in water. I constructed a primitive raft out of doors and timbers, on which I could transport a few people, including the mayor, over the millrace to the bank at the edge of the town. Frau Löffler met a tragic end. She'd already got two of her children up to the attic and went back downstairs to fetch her six-year-old son. While she was in the living room, the water rose so rapidly that, although she was a good swimmer, she and the boy were no longer able to get under the doorframe. Both of them drowned in their living room.

A dreadful fate also awaited some fifty people, who suffered a ghastly end when, hearing the air-raid warning, went down into the deep bunker at their works. The growing pressure of the rising water was such that they couldn't get the door open, and they drowned.

Frau Tigges

RESIDENT IN WICKEDE

That night, around two, I was woken by the terrible storm and the din. I ran to the window and screamed, 'Water! Water!' I roused the children from their sleep and ran up to the attic with them. From the roof light I saw a number of buildings carried by on the house-high waves. Suddenly the neighbouring house rose up and twisted off its foundations, bounced against ours and tore half of ours away. The next-door house was carried a hundred metres down towards the valley and broke up with a crash, tossing the neighbours who were in it into the floodwaters. Anxious hours ticked by. In the grey light of morning, a low-flying aircraft circled over the Ruhr Valley. Only around midday could relief workers get through to us and take us to safety.

Herr Clemens Mols

POSTMASTER IN WICKEDE

I was with my wife on the way from Wiehagen to Wickede/Ruhr when the horn started. Back at home at the post office, my wife, who was strangely restless, asked me to listen and find out where the English planes were. I could get this from a warning line connected to the post office. The report ran as follows: 'Enemy aircraft flying low above Arnsberg and the Möhne Lake.' As Arnsberg lies not far off from Wickede, my wife went to wake up the people in the house next door. While she was doing this, I stood in the open window on the first floor with the view of the Möhne Lake. The humming of the planes came from a distance.

Suddenly we heard an unusually loud detonation and in the direction of the Möhne Lake I saw a high column of water or smoke soaring up.

The night was so clear that one could see this very distinctly. It could only have been caused by bombs – the more so as some other smaller detonations followed. Then all was quiet. Further reports said, 'Enemy aircraft low over the Eder Lake.' The Möhne Lake was not mentioned any more. Calmed by this, I thought the danger was over. I went down to the cellar of the house next door where the inhabitants – mostly women and children – had gathered. They were talking and laughing, and I told them, 'I believe it's over, so we can go to bed.' On my wife's request, however, I went up again to listen to what the situation was.

As I entered the post office rooms I heard the telephone ringing. I picked up the receiver and heard the familiar voice of the attendant of the post office in Arnsberg, who said: 'What, Herr Mols, are you still at the post office? The Möhne Dam is broken – the waters must have reached Vosswinkel by now!' (This was about 5 km away.) At first I was giddy with fear, but then I rushed back at once to the neighbours' cellar, calling, 'Get out at once, all of you, to the upper village. The Möhne Dam is broken and the water has reached Vosswinkel.' In the meantime the electric light went out. I said to my wife, 'You go along. I'll stay here and try to wake people by phone.' She refused to leave me alone, I begged her, 'Hurry to Frau Brunberg and wake her up so she can escape with her three children – and tell Fraulein Wilmes.' It was a house very near where there were no men, as they were away in the army.

While my wife was doing this, I tried to wake people by phone – but without success. They all seemed to be asleep. In the meantime, my wife came back and insisted we go to the upper village. When we had gone some fifteen to twenty paces, the air became like a cold, damp screen of fog. 'You cannot run – the water will be here in no time,' I said to my wife. 'We must go back to the house.'

We were scarcely back inside, when the water caught us up in the hall. We slammed the front door and hurried upstairs. We saw the water rushing into the house with a terrific speed, as a sulphurous cloud of vapour mounted from the cellar – it was a short-circuit from the battery stored there.

I tried to comfort my wife, who wailed as she wept and prayed – these were terrible hours. I was working out how to get on the roof of the neighbouring house in case ours was swept away in the flood – the other house was taller and had a flat roof. We carried our bedding, clothes and linen up to the loft – we couldn't tell what height the water would reach. I was running from one window to another to see if the walls were still holding up.

Above all this terror, the moon was shining brightly and was reflected on the water. An enemy aircraft was flying low, up and down the river. It kept very low, and we wondered, 'Is it going to drop bombs?' but nothing happened – it passed. All the houses around were rising above the water with just their upper storeys visible – smaller buildings had disappeared in the current.

I decided that the water was no longer rising – and I called out to my wife, 'The water has already gone down, and it's receding fast.' She wouldn't believe it – I had to show her. Outside the day was dawning and the water continued to recede. On the ground floor there remained a muddy deposit about 30 to 40 cm deep. We stepped through the mire and tried to open the back door of the house, but the pressure of the water had jammed it in and only with difficulty and the aid of other people could we eventually get it open. Outside the water was still a metre deep and it would be hours before it had receded completely. Everywhere we could see scenes of terror and destruction. A metre-deep layer of the railway embankment was washed away, a heavy engine was completely carried away, rails 100 metres long and more were transported into the meadow. Many houses were swept away, along with their inhabitants. Two big factories were completely destroyed and could not get back into operation.

Before the war, hard-working people had lived here, cultivating fertile fields. Even two years later the land was still a wasteland.

Hans Werner Konig

HEAD OF THE ASSOCIATION OF RUHR RESERVOIRS

I came from Balve five hours after the raid. I stopped the car and realised there was no water. I realised that something terrible must have happened. There was a big hole in the dam wall.

Vikar Kemper

RESIDENT CURATE IN HIMMELPFORTEN

After one or two days, once the waters had dispersed, we spoke to people and asked where the pastor of Himmelpforten was. They said, 'In the air-raid shelter! That's certain. And his sister who came to visit of an evening is there too – likewise his housekeeper.' The latter was washed ashore the following morning in Neheim and identified. But where was the pastor? He would have been in his air-raid shelter. Only some time later was it possible for a handful of men from the church council to dig and hack away the rubble. We dug down about a metre. The air-raid wardens said that it was pointless. But the men from the church council wouldn't have it, and said, 'We'll dig on, right down into the earth.' We dug with our hands – we found a foot, then a black stocking and black trousers. And there was Pastor Joseph Berkenkopf, who'd lived here almost three decades doing his ecclesiastical work. The poor man had been overtaken by a gigantic wave of water and scree. He lay aslant the space, his head and face down. It was terribly difficult to drag him out. Eventually we got him out and laid him on an old door that we found nearby. Then, for me as a priest, there was a defining moment – something joyful. A man from the church council said, 'Hats off for a prayer.' And everyone stood, still and reverent – including the policemen from the Essen air-raid defence who very possibly thought quite differently from us.

Max Freiherr von Boeselager

Resident in Wickede

As soon as the receding waters cleared from the road into the valley I set out on a recce trip on my bike. On the railway embankment and on the hillside in the wood were groups of frightened cattle. It was extraordinary that so many had managed to escape. Instinctively the animals had sensed the approaching danger. Warned by the bellowing of the cattle down in the valley, many succeeded in escaping just in time. Thus the first alarm signal in Vosswinkel was a herd of horses which came into the town at a gallop from the valley, just before the waves rushed in below.

When the floodwaters receded to the river bed, they left the fertile valley a wasteland. Meadows, pastures and fields were washed away or inundated with banks of sand or shingle. The railway embankment was washed away in parts and torn up, the rails were bent like wires and mangled together. Everywhere you saw the carcasses of animals caught up, and here and there one's gaze fell on a drowned body. Eerily, as if in supplication, a child's arm with an outstretched hand protruded from the mud.

Wickede had been hit badly. No warning came through here. Most people were asleep in their beds when the floods burst through. The list of the dead was long.

Franz-Josef Cloer

The whole community was horrified and in a state of shock. Once the water level started to go down, over the next couple of days, we could begin to see the extent of the catastrophe. The whole of the Möhne Valley was flooded. Where before there had been houses, you could see only the remains of the foundations. The houses had simply disappeared – the water had carried them away. The Ruhr Valley was covered with carcasses of cows, sheep, chickens, horses and all kinds of animals, as well as a great many bodies. For weeks and months afterwards, bodies were being found and identified – although some of the bodies were in such a state that they were beyond recognition.

Victims of the Eder flooding

Christian Kohl

LOCAL HISTORIAN IN HEMFURTH

16 May was a Sunday with brilliant sunshine – and moreover, it was Mothering Sunday. The Eder reservoir was filled to the level of the top of the barrier and was flowing over. The water poured in a white mist over along the length of the wall. A picture of profoundest peace. Only the watchmen and the Hanover-Münden officials, who patrolled along the wall, their carbines slung over their shoulders, reminded us that we were at war. Even the 2-cm flak guns, until recently sited along the wall, had been removed a week before.

About 1.15 am I was woken by the noise of low-flying aircraft. My family and the others living in the house went down to the cellar. I went outside with a friend to investigate. In the bright light of the full moon we could make out the surrounding area clearly. Repeatedly, low-flying four-engined aircraft approached from the northwest. On one aircraft I could see the pilot had

Schalber/Bundesarchiv Bild 101I-637-4194-13A.

'Everywhere we could see the scenes of terror and destruction . . . a heavy engine was completely carried away, rails 100 metres long were transported into the meadows.' Herr Clemens Mols

opened the window of his cockpit and had stuck his head out to look around. Then I heard two explosions, one after the other – without anything happening. According to the account from the observation post which stood right on the wall, one of the aircraft set off a phosphorous flare bomb right between the power stations on the open side of the wall as a marker, and it bathed the area in an eerie light.

Shortly after 2 am, another aircraft flew very low from the direction of Waldeck Castle towards the wall and set off a third explosion, which was immediately followed by a terrible roaring, as though the wood was being whipped by a raging hurricane. I rushed into the air-raid shelter in my house and shouted, 'The wall's broken! Get out – get out to higher ground!' As everyone ran for their lives, a massive wave rolled towards us down the valley. We'd hardly got up the hillside to safety when the surrounding houses were engulfed by water. With a monstrous cracking, the great iron construction of the Messhänge Bridge, situated in the upper part of Hemfurth, was ripped up.

In the stables, the drowning cattle bellowed terribly into the night. Farmhouses in which there were still people, tumbled into the swirling mass of water. No-one could do anything to help.

Wilfried Albrecht

Age fourteen and working as an apprentice in Hemfurth power station

I was still lying in bed when the bombs fell, since up until then there'd been no night alarm – so we didn't worry any more about it. We had always assumed that enemy aircraft were gathering over the Eder Dam before flying on for an attack on some more distant target. However, after the first explosion, my mother shook me awake and I ran from our wood-built house into the street. From there I could see the water from the lake gushing through the hole in the wall.

Motor boats and whole boathouses were carried through the breach in the dam and down into the valley on the huge waterfall. Thirty years after the catastrophe, people doing dredging work for a hydro storage facility in the former bed of the Eder, found some strange floats whose provenance they couldn't explain. But I remembered immediately that these had belonged to the boathouses which had been drawn through the breach in the wall by suction. The power of the enormous explosion raised the glassy surface of the lake close to the wall metres into the air. The tidal wave threw several sailing boats and rowing boats high on to the shore. Several areas of the bank were also badly damaged by the waters.

The first deaths in the Eder disaster were in the Hemfurth I power station. The power station was manned every night by four people – an engineer in each station and two men on the shift on watch. I was also part of the Eder reservoir staff, and had begun my apprenticeship four weeks before, but that particular night I wasn't on duty. When the aircraft attacked, the engineer in power station I tried to get down into the cellar under the turbines as shelter against the bombs. He must have believed that he'd be safest there from bomb splinters. No-one could have reckoned that the wall could ever be destroyed. Down in the power station the engineer must have been overtaken by the deadly waters. He could never have survived. We looked for him for ages later, but couldn't find him. Only months later was his corpse found by clearance workers, over twenty kilometres away in the Wabern area.

Heinz Sölzer

RESIDENT IN AFFOLDERN, JUST DOWN THE VALLEY FROM HEMFURTH

My parents and I were woken that night by the sound of low-flying aircraft. As a kid of fourteen, I was naturally curious, and ran out on to the street to look at them. You could make out the bombers with their orientation lights on. After a while I saw a white cloud of mist rise over the wooded mountainside in the direction of the dam, and felt a thundering impact. Shortly after that our phone rang. It was Mayor Ochse from Hemfurth, and he said frantically that the dam had been hit and the water was gushing out, and that we had to alert all the residents in the village of Affoldern.

I quickly grabbed my bike and rode down into the village to spread the alarm. Along with a friend I bumped into on the way, we knocked on lots of windows, and explained and shouted, 'The dam's burst. Save yourselves – get up the hill!' Several times the chain came loose on my bike and delayed my own escape back up the hill. Already the water was roaring in with such a speed that I could never have outrun it on a bike. A number of houses were simply dashed out of the way and thrown around by the power of the water, collapsing with dull crashes.

Elise Schäfer

It was Mothering Sunday, and I went walking with my mother to the churchyard and we marvelled at the wonderful burgeoning of nature around us. My father said later at dinner, 'We're in for an autumn harvest the like of which we've never seen.' That night, around two o'clock, we were woken by the planes and I looked out of the window. Since we lived at the end of the village, I could see the aircraft clearly as they circled over the area of Buhlen Station. I hadn't given a thought to the Eder Dam.

My mother called out from downstairs, 'Children, get inside, the fliers tonight really mean business.' My sister-in-law came running into my room in a state. I hesitated a little in getting back. Suddenly there was such a massive tremor that the door rattled. With that I ran at once to my grandfather – about 86 at the time – who was asleep upstairs. I said, 'Quick, get up – the planes are here.' 'Oh child,' he said, 'they won't do me any harm.' So I ran back downstairs to my mother.

My father peered out of the window and said, 'That's not a smoke cloud there in the distance. It must be something else.' I went out with him into the street, where people were running towards us from out of a public air-raid shelter, and talking frantically among themselves. In the distance we could hear rushing water and we thought they were letting water out from a reserve levelling basin

Suddenly, from in the village above there was a great clamour. Two schoolboys were spreading the alarm which the Mayor of Hemfurth had raised by phone. I think it was Heinz Sölzer and Heinz Heck who'd cycled into the centre of the village and were shouting, 'Save yourselves. The water's coming.' We didn't know what to think until a soldier, who came from a small power station in the valley, saw me and shouted, 'Is everyone here away? The water's coming!'

'Down here no-one's gone,' I said to him, and turned and ran back.

It occurred to me that Grandpa was still upstairs. I ran up again and shouted, 'Grandpa, quick – out of bed. The dam's broken.' He got up at once. When I tried to turn the light on nothing happened. I quickly packed a few clothes into a small case, expecting that my father would have harnessed up and I could take a few more things with me in the wagon. When I looked down into the yard there was no-one there. I ran back to Grandpa, who was standing there in his shirtsleeves.

At that moment the water came hurtling down the street. I thought, 'We've only got a moment – the water's rising like high tide in the Eder Lake.' I ran back up the stairs inside. The animals in the stable below were bellowing terribly. On the stairs I met my grandfather and said, 'Grandpa, we can't get away. We'll have to stay here.' He asked, 'But where are the others?' 'I don't know' – and we climbed up the stairs.

My grandfather sat down on a sack of vegetables by the chimney. I sent up a lot of little prayers to heaven. Downstairs there was a roaring through the houses as if the water were going to consume everything. Occasionally we heard the gushing water get louder every time a house or barn went under. Suddenly there was a terrible impact, such that I feared that our house would go up, but it had been our neighbour's barn, which had been buffeted against our house. It was packed with straw and animal fodder.

From then on it got calmer in the house and the roaring receded. In the height of my fear of dying, I'd taken a big baker's basin from near the chimney, with a view to surviving by floating in it. In climbing into the basin, a Bible story came to mind: '. . . and they put Moses in a reed basket in a sheltered spot on the Nile so he would be saved – and he was saved'. That's what I hoped – to be able to survive on the baker's basin. Divine providence allowed me to live. The water didn't rise up as far as us.

As morning drew on and I got out of my basin, I looked out of the window and saw a water wilderness all round. Trembling, I went downstairs and saw the devastation in all the rooms. In the stable, Rosa, our mare, wanted to come to me through the broken door, but couldn't because of the rubble. Even her little foal, just four weeks old, was still alive. It must have held on to his mother as she floated in the flood.

August Kötter

GERMAN SOLDIER ON LEAVE FROM THE RUSSIAN FRONT, AT HOME IN MEHLEN, A LITTLE FURTHER DOWN THE VALLEY FROM AFFOLDERN

There was a great to-do outside and people were shouting, 'The dam's bust! Get away, up the hill.' I quickly gathered a few vital things together and said to my wife, 'If the water comes, we've got that superb pig in the stall – it would be dreadful if it drowned. We've got to see if we can get it upstairs.' I rushed down to the stall and tried to drive the pig upstairs via the cellar steps. The animal ran up the stairs as though it had done it a hundred times before. It was pure instinct – the animal must have sensed the danger.

That moment my wife came rushing in, 'Come quickly – the water's rising.'

'Right,' I said, 'if the water's here already, then we can't get away and we'll have to stay put. It's too far to get to safety in the hills.'

To orient myself, I glanced out of the window to the valley, in the direction of the dam. I saw water coming like an avalanche. The water was steep – as if it had been cut through – like a straight wall with foam rushing before it. So what were we to do now? I said to my wife, 'We must wait and see what happens.' I didn't want to show my fear in front of her.

Luckily, the children had escaped with the other people, as the next moment there was a loud report and the front door crashed in. I was up to my chest in water and only with difficulty could I pull myself up by the banisters. The flood pushed the kitchen door open, where the pig was squealing loudly. The next wave of pressure ripped the kitchen window outwards and the pig was driven bellowing out of the window with the flow of water.

We fled up another storey, but the water came still higher and higher, until we eventually arrived at the attic, immediately under the roof. There we stood and listened to the dreadful roaring. I pulled a few roof tiles up so I could see what was happening. To my horror, the next-door house wasn't there any more, just like a number of other houses in the neighbourhood. Then there was a crash in the house and lumps of debris smashed into the walls and destroyed the entrance. All of one side crashed in along with half the kitchen.

That very moment, as I was looking down, there was a terrible tremor. I leapt back up the stairs, as I'd gone down a step or two, and grabbed my wife. I barely had her in my grasp when the whole front of the house collapsed with an almighty roar. We rushed back up to the attic and listened for every new sound. Still more chunks of masonry broke off our house and disappeared with a crash into the water. In between, we heard a distant dull crash, combined with gurgling noises, as another house or barn in the village went under.

By now, our house was half demolished. But I saw the house behind ours was still undamaged, because ours was standing in front protecting it. So I started to think how we might get over to it. Then I got the idea to use the wooden attic ladder as a raft to get across the calmer water behind our house. I still don't know how I managed to wrench the ladder loose.

Before we climbed on to it, we bound the ladder to a roof beam with a washing line so it couldn't be swept away. First of all my wife had to tie herself to this rope. I told her she should let herself slide down as slowly as possible, but she couldn't manage it and she plunged down into the water with some force. The cord ripped through her hands, skinning the skin off until her palms until they bled.

To steer our life-raft, I'd taken down the flagpole we always used for the flag on the Führer's birthday. Eventually I climbed down the rope and tried to climb on to our ladder raft, which was now floating below the surface of the water. My wife was sitting on it, and her weight had half submerged it. Then it suddenly capsized and threw her into the water. Without hesitating I dived in and was able to catch her – but it was a tough job to pull her, although luckily she was still attached to the washing line.

My wife was now rigid with fear and it took a long time before I could push her back, like a wet sack, through the bathroom window into our house. As she pulled herself together, she became terribly thirsty. I told her, 'You certainly can't drink this contaminated water,' and she replied, 'There's still a flask of milk in the pantry.' I climbed out again through the bathroom window and in through the kitchen and saw the milk flask, which was swirling around in the eddying water. I quickly fished it out and she drank it all down in a single draught.

Schalber/Bundesarchiv Bild 101I-637-4193-17.

'Once the water level started to go down, over the next couple of days, we began to see the extent of the catastrophe.' Franz-Josef Cloer

After a few hours the water receded and we heard voices in the distance. They were trying to get help to us with a boat, but they failed because a powerful maelstrom had formed in front of the house. Eventually a pioneer boat from Hanover-Münden became aware of us, alerted by the loud shouts of the people on the slopes of the valley.

The rushing water had reached a height of five metres around our house. I thought then, 'At the Russian front, in the fighting and the cold, you were lucky – and here, in your homeland, you were almost horribly drowned.'

Even our pig, that I'd taken so much trouble over, was rescued. A few days after the flood we searched down the valley and found him about ten kilometres away in a stall where a farmer had temporarily taken him in. When my wife called out to him, 'Come! Come!' he ran over to her, squealing happily. The pig had recognised the voice of the one who fed him!

Squadron Leader Jerry Fray

SPITFIRE PILOT, 542 SQUADRON, PHOTO RECONNAISSANCE

I left RAF Benson at 07.30 hrs – visibility was exceptional. When I was about 150 miles from the Möhne Dam, I could see the industrial haze over the Ruhr area and what appeared to be cloud to the east. On flying closer I saw that what had seemed to be cloud was the sun shining on the floodwaters. I was flying at 30,000 feet and I looked down into the valley, which had seemed so peaceful three days before, but now it was a wide torrent with the sun shining on it. Twenty-five miles from the Ruhr, the whole valley of the river was inundated, with only patches of high ground and church steeples which I had seen as part of the pattern of the landscape a few days before, showing above the flood waters. The even flow was broken as it rushed past these obstacles. As I came nearer the dam, I could see that the water was about a mile wide. I was overcome by the immensity of it, and when I realised what had happened I just wondered if the powers that be realised just how much damage had been done.

The Ruhr was covered with haze and when I broke clear of this, I began my photography, moving up towards the dam. It was easy to pinpoint because the breach showed up, and I could see the water rushing through. The control house at the foot of the dam, which I had seen two days before, had already disappeared. The level of the water above the dam had fallen, leaving huge tracts of dark brown mud around the edges. This was eight hours after the bombing. The upper reaches of the lake were completely dry, except for a small portion where the sluice gates had been closed.

I then flew on to the Eder Dam. The floods were easy to see. The long winding lake above the Eder Dam was almost drained and as a landmark it was no longer there. If it hadn't been for the floodwater breaking out of the breach in the dam it would have been difficult to find the lake. The water flowed through the narrow valley and from 30,000 feet I could see the course of the original stream. It stretched eastwards and northwards to Kassel.

Joseph Goebbels

GERMANY'S MINISTER FOR PROPAGANDA

Diary entry for 20 May 1943

In the evening Speer phoned me and gave me a report about the dam disaster. He flew out there immediately and saw for himself the damage sustained. Thank God it's not as bad as even Speer had suspected. He set in place a whole lot of measures – most particularly he brought back a large labour contingent from the Atlantic Wall and set them to work repairing the damage. Overall it was a matter of six to seven thousand men, who were already en route and would soon be on the spot. Speer gave the optimistic impression that by the start of the next week he would be able to return production to fifty per cent. By the end of the next week they should be running on full production again. Speer is an organisational genius. He doesn't let himself be put out by serious setbacks, and in returning the equilibrium he takes the measures required and doesn't shrink from dictatorial rulings when the situation calls for it. The Führer has given him absolute power and he has implemented it most thoroughly.

Albert Speer

GERMAN MINISTER OF ARMAMENT AND WAR PRODUCTION

(From his book, *Inside the Third Reich*)
The report that reached me in the early hours of the morning was most alarming. The largest of the dams – the Möhne Dam, had been shattered and the reservoir emptied. As yet there were no reports on the three other dams. At dawn we landed at Werl Sirfield, having first surveyed the scene of devastation from above. The power plant at the foot of the shattered dam looked as if it had been erased, along with its heavy turbines.

A torrent of water had flooded the Ruhr Valley. That had the seemingly insignificant but grave consequence that the electrical installations at the pumping stations were soaked and muddied, so that industry was brought to a standstill, and the water supply of the population imperilled. My report on the situation, which I soon afterward delivered at the Führer's headquarters, made a deep impression on the Führer. He kept the documents with him.

The British had not succeeded, however, in destroying the three other reservoirs. Had they done so, the Ruhr Valley would have been almost completely deprived of water in the coming summer months. At the largest of the reservoirs, the Sorpe Valley reservoir, they did achieve a direct hit on the centre of the dam. I inspected it that same day. Fortunately the bomb hole was slightly higher than the water level. Just a few inches lower – a small brook would have been transformed into a raging river which would have swept away the stone and earthen dam. That night, employing just a few bombers, the British came close to a success which would have been greater than anything they had achieved hitherto with a commitment of thousands of bombers. But they made a single mistake which puzzles me to this day. They divided their forces and that same night destroyed the Eder Valley dam, although it had nothing whatsoever to do with the supply of water to the Ruhr.

A few days after this attack, seven thousand men whom I had ordered shifted from the Atlantic Wall to the Möhne and Eder areas, were hard at work repairing the dams. On 23 September 1943, in the nick of time before the beginning of the rains, the breach in the Möhne Dam was closed. We were thus able to collect the precipitation of the late autumn and winter of 1943 for the needs of the following summer. While we were engaged in rebuilding, the British air force missed its second chance. A few bombs would have produced cave-ins at the exposed building sites, and a few fire bombs could have set the wooden scaffolding blazing.

CHAPTER 11

Aftermath

From the time when the aircraft took off, the atmosphere at Scampton was tense. Two aircraft which had been forced to turn back before reaching the targets landed between 12.30 and 1.00, but the first of the Lancasters which had survived the raid on the dams started arriving home from 2.30 on the morning of 17 May.

Flight Lieutenant Harry Humphries

ADJUTANT, 617 SQUADRON

The time passed very slowly. Occasionally the silence in the anteroom would be broken by a stifled cough, or whispered conversations. Why the hell people had to whisper I don't know, but then I found myself doing it every time I addressed Fay Gillon, the intelligence officer, or anyone else. I kept nodding off and suddenly I imagined I heard the sound of an aeroplane. I must have been dreaming – no, it wasn't a dream. The drone came nearer. Everyone in the room was listening.

Aircraftwoman Morfydd Gronland

We stood silently until the final sounds of their engines died away. Then we drifted away to our duties.

There was no sleep for anyone that night – our hearts and minds were in those planes. We WAAFs just sat waiting. We had laid out the tables and a hot meal would be ready on their return.

The night wore on. Twice we heard the roar of engines and rushed outside to see what planes were returning. One was AJ-W, piloted by Flight Lieutenant Munro. The other was AJ-H, piloted by Pilot Officer Rice. They had returned early. Munro's plane had been badly damaged by flak over the Dutch coast and had to return. Rice's had flown too low over an inland sea in Holland, and hit the water, tearing the bomb from the aircraft. By a miracle the plane did not crash and the pilot by superb skill brought his crew back to Scampton safely.

We settled back once more to wait. The WAAF Sergeant made us all coffee and calmed us down. 'It will not be long before our boys start to come back,' she said. We nodded back at her.

Sergeant Jim Heveron

IN CHARGE OF THE ORDERLY ROOM

When the ops room told us the aircraft had started to cross the English coast on their way back, we all went out to dispersal and counted them back. It was obvious that several were missing, and what little the aircrews could tell us made us certain that this had been a really rough operation. Finally, about 4 am, it became obvious that no more would be coming back.

Those of us who were close to the crews were in a state of bewilderment for some days. I had become a trifle hardened to crews who I knew would not be returning from operations, but one never grew entirely immune from the sense of loss of fellows one had known intimately.

Flight Lieutenant Harry Humphries

ADJUTANT, 617 SQUADRON

It was fast approaching the time for the return of the bomber force. Someone had the bright idea of switching on the wireless and tuning into the shortwave for the purpose of contacting our returning aircraft. It worked sometimes, and sometimes the reception was entirely unintelligible. On this occasion it actually did function quite well, for as soon as we heard the first machine circling the aerodrome, we picked up quite clearly the pilot's request to land. I am not sure now, but I think it was Squadron Leader Maltby.

Ruth Ive

WAAF, RADIO OPERATOR

When dawn broke, planes started to come back again – but in a terrible state. When I went out from the Nissen hut, all the fire engines and the ambulances were on the ground. And the planes, they were shot up and the undercarriages were only half extended; they were in a terrible mess. I asked where had they been to, and was told the German dams. I have never forgotten the state of those planes which had staggered home, all through the very early hours of the morning.

Arriving back

With a profound sense of relief, eleven Lancasters landed back at Scampton – some unscathed and others bearing the scars of flak and tracer fire. Unaware of the fate of the eight other aircraft and their crews, there was a sense of elation among the men – then the cost of the mission became apparent.

Harry Humphries

ADJUTANT, 617 SQUADRON

I said to Maltby, 'Hello Dave, and how did it go?'

'Marvellous,' he said, 'absolutely marvellous.' He had never seen anything like it. 'Water, water everywhere – wonderful, wonderful.'

Maltby asked me if I'd had Nigger buried for the Wingco. 'It was just worrying Gibby, I know. It struck me that superstition means nothing anyway, even though I always take this hat with me.'

'This hat' was David's field service or 'fore and aft'. It was a filthy thing, covered in oil and grease, but he would not be separated from it, even on parade.

'Well, see you later over a beer,' he said, and shouted to the rest of his crew, 'so long sprogs. Thanks for coming!'

Squadron Leader Maltby addressed his crew as 'sprogs' or 'rookies' because he was the only operationally experienced member of the crew, but this night his boys had gained their spurs.

Sergeant Basil Feneron

FLIGHT ENGINEER, AJ-F

When we landed, I jumped out and the first thing I did was kiss the ground – it was something I always did after an op. We didn't know we'd been hit until then. Even Hughie Hewstone, the wireless operator, hadn't seen the jagged hole about two feet square, just behind him. The shell narrowly missed the aileron controls.

We went straight to debriefing and a cup of tea. Bomber Harris said, 'Well done, lads.' There was an atmosphere of congratulations

– good show, well done. After that we went off for a meal of bacon and eggs. It tasted especially good that morning.

Sergeant George 'Johnny' Johnson

BOMB-AIMER, AJ-T

Even with a burst tyre, Joe controlled the aircraft very well. It had been exacting – but this was very much a crew effort – it was always a crew effort.

Flight Sergeant Leonard Sumpter

BOMB-AIMER, AJ-L

We landed and got out of the plane – we were pretty tired with all the concentration. We all felt a bit drained – and we congratulated each other, and we were all very cheerful. We went back to the debriefing room, sat down and told them what had happened – and had a cup of coffee.

There was always a girl there when you came back, with an urn of coffee – she just sat there, keeping it hot for when you got back.

Flight Sergeant George Chalmers

WIRELESS OPERATOR, AJ-O

I think I was always relieved when I got back! You just wanted to get your feet back on the ground again. When we landed we were debriefed, and then we were allowed to go back to our beds – have a bit of a rest. After debriefing I went straight to bed after having my breakfast, then I got up at about one o'clockish. However, I believe certain people did celebrate rather a lot. It wasn't until later that morning – about lunchtime – that we started to hear different tales from various people about what had happened, and we saw the press announcements too. We learned more about what happened the following day than when it was actually happening around us.

Flight Sergeant Ken Brown

Pilot, AJ-F

When you got back – well, you've never had a bowel movement as great. You just felt haaaa – we'd made it. We didn't notice, really, until we got out of the aircraft, that a number of the ground crew – in fact, all of them – were crying, and it was rather disturbing. We were rather elated that we'd made it back, and there was this bunch of people crying. I don't think it hit me until Flight Sergeant 'Chiefy' Powell came in, and the tears were just pouring down his face. We looked at him and said, 'Chiefy, what's wrong?' And he said, 'Have you any idea where any of the other boys are?'

Out of the nineteen, one flown by Officer Rice had hit the water on the Dutch coast. As a matter of fact, the tail-gunner was under the water when the aircraft pulled out. He made it back. Another aircraft, flown by Flight Lieutenant Munro, was hit by flak and his intercom went. So consequently, of the nineteen you're now looking at seventeen that went beyond the Dutch coast.

The reception at Scampton

Air Force top brass and ground crews who had waited anxiously throughout the night were out in force to welcome the returning crews and congratulate them on their success.

IWM CH9683.

After the raid Gibson's crew are debriefed by the Intelligence Officer (in glasses). Standing behind are Air Chief Marshal Sir Arthur Harris (left) and the Hon. Ralph Cochrane (right). Seated, left to right, Spafford, Taerum and Trevor-Roper.

Flight Sergeant George Chalmers

WIRELESS OPERATOR, AJ-O

I was first out of the aircraft to be met by Air Chief Marshal Harris, Air Vice-Marshal Cochrane and Group Captain Whitworth, and at the shock of seeing them I nearly fell over in shaking their hands. They congratulated me on my Morse, which was easily read by them.

Sergeant Dudley Heal

NAVIGATOR, AJ-F

We flew back over the North Sea and landed back at base, to be met by the chief of Bomber Command, Sir Arthur Harris, Group Captain Whitworth, Air Commodore Cochrane. They came to meet the aircraft so we realised then how important this operation had been considered. They seemed very pleased with what we said and we went and had a proper full debriefing before getting some breakfast.

We asked at once about the others. There were still eight to come back so we sat there, drinking tea and waiting for them. But as time went on we realised that they weren't coming back because they would have run out of petrol. We learned eventually that we lost eight aircraft out of the nineteen that started. We didn't know what happened to any of them at that time, although we'd seen two go down. We learned that some went down over the Eder and then one over the Möhne. One, we learned subsequently, Squadron Leader Young, was shot down at the point over the Dutch coast where we had very nearly bought it, and so there were gun crews waiting. Having shot down one aircraft, they were fairly confident of doing the same with us, but thanks to Ken, we got away with it.

Kenneth Lucas

Ground crew

I was naturally fast asleep – and I think I was still asleep when they came back. Then Gibson had the whole squadron together and told us what had happened, and then as soon as the damaged planes were removed, we went on leave.

There was much sadness – they were very brave men and the cream of the RAF, and I felt humble by comparison. I don't feel that ground crew have been neglected – we had a job to do – it was necessary and they did rely on serviceable aircraft – or they didn't reach the target.

Gibson told us what had happened, and we thought it was great. There was a feeling of elation – but at the same time we thought of the crew who had lost their lives – who didn't return. Later on, when I saw a film of what had happened, and saw animals and so on floating in the water, I thought I'd been a little part of the cause of that, and that hit me. Gibson told us he had a slight feeling of guilt when he saw a car in the water, and the headlights suddenly went out, and they realised what had happened.

Flying Officer Edward Johnson

Bomb-aimer, AJ-N

When we got back there was a fairly comprehensive debriefing – a fair bit of excitement that we'd been successful in doing at least two of the dams. We didn't know much about the others at that time, because the waves that were to attack the Sorpe and the Ennepe, and one or two other dams didn't go off until after us. As it turned out, they didn't do an awful lot of damage. They were a different type of dam – not the same construction as the Möhne and the Eder, so I don't think they could have been expected to do too much damage with that type of bomb.

After debriefing we didn't go to bed – there was a good deal of boozing went on through the night. I haven't the faintest idea what time I went to bed.

John Elliott

NCO GROUND CREW

G-George was not in a bad state when it came back, strangely enough. Not bad at all, considering what Gibson had done. He must have been extremely lucky – to do what he did, I would have thought he'd have been blasted out of the sky. He hung about that target for so long – I think he was an extremely fortunate man.

Gibson came back and all his engines were all right – pretty well everything was all right and none of his crew was injured.

The crew were certainly relieved to *be* back – and I think probably a little bit amazed at what they had done. When we realised what they had done we were amazed as well.

On landing they were certainly not bouncing with joy – they must have been awfully tired. It was a question of getting to the debriefing, having a meal and some sleep if it was possible. But I doubt they'd have got much sleep because they knew there were still quite a few aircraft missing. I think they were probably up most of the night in the mess, waiting to find out who was eventually going to get home. But there were quite a lot that didn't.

Flight Sergeant Ken Brown

PILOT, AJ-F

How the ground crew worked on our aircraft – and believe me they did! My aircraft was so badly damaged it had to go back to the factory. I have a great deal of respect for the ground crew personnel. They did a tremendous job and it's unfortunate at times we don't really recognise what they did. They were really a godsend to us all.

Flight Lieutenant David Shannon

PILOT, AJ-N

The atmosphere was one of elation that it had been so successful – but also of depression that we had lost almost fifty per cent of the squadron – of the people who took off – because of the twenty-one crews chosen for the original squadron, nineteen crews took off on the night. There were two early returns. Of the seventeen that proceeded to the target, eight were lost – so it was very expensive in crews – but I would say that with the success of the raid, it was worth it. Even if we had lost everybody, and it had been a success, it was a tremendous blow and undoubtedly retarded the German war-effort considerably.

Aircraftwoman Olwen Jones

WAITRESS IN SCAMPTON OFFICERS' MESS

On the morning after the raid, all hell was let loose. The bar had been open since the aircrews returned. Pilot Dave Shannon had fallen and blacked his eye and no-one wanted breakfast. It really hit us when they were talking about how many hadn't come back. I will always be proud to have known such brave men.

The cost of the raid

At Scampton those waiting for the aircraft to return – and those who had already made it back – watched the skies. The men compared notes from their mission about the aircraft they had seen go down and it soon became apparent that eight crews would not be coming home.

Aircraftwoman Morfydd Gronland

Some time later we heard the sound of engines in the far distance. Once again we all ran to the landing strips. The first planes came in low and taxied to a halt. Then at irregular intervals other planes began to land. We were ordered back to the Sergeants' Mess to start serving the first arrivals. We waited but no aircrew came in. Two hours later our WAAF Sergeant entered, she called us together. 'I must tell you now the very sad news. Of our nineteen aircraft, only eleven have returned. Eight have been lost and fifty-six of our young boys will never return.'

We all burst into tears. We looked around the Aircrew's Mess – the tables we had so hopefully laid out for the safe return of our comrades looked empty and pathetic. The Sergeant told us to go to our quarters and try to get a few hours' sleep, because tomorrow would be another working day.

The following few days were a nightmare. We were still shattered by the terrible loss, but gradually we began to adjust the squadron routine. However, things would never ever be the same again.

Flight Lieutenant Harry Humphries

ADJUTANT, 617 SQUADRON

I saw more of the crews in, and even though they were in high spirits, they confirmed that we had taken a beating as regards casualties. I decided to walk over to the watch office – the flying control tower. My worst possible estimation was insignificant compared with the shock I received. Eight blanks on the blackboard. It was hard to accept.

Flight Sergeant Leonard Sumpter

BOMB-AIMER, AJ-L

The losses were the down side – some of them were friends with other crews, but it wasn't so bad in the Sergeants' Mess, because it was expected in the air force that the chap next to you – you may simply not see them any more. You go out and hope that you all get back. And they had a real knees-up in the Officers' Mess that night and everybody was really tipsy. We had a good do in the Sergeants' Mess too. But as I say, you got in the plane and took off and it was just fate if you came back or not. Our crew used to call me Satan – I used to tell them I was too wicked to die. I said, 'You'll never go down with me in the plane' – and that was a big joke.

'We'll be all right, we've got Len with us!'

John Elliott

NCO GROUND CREW

It was a shocking thing when we realised who hadn't come back. They were people we had been working with for so long and seen so much of. We probably saw more of the aircrews there than we would have on a normal squadron, because they were training day and night and you would have your crews coming out pretty well all through the day. We got to know them quite well and it was a bit of a shocker to lose so many people.

Flight Sergeant George Chalmers

WIRELESS OPERATOR, AJ-O

There was an air of despondency. People weren't sitting where they use to sit – and never would again. The press, however, were

IWM CH9936.

Twelve men of the RAAF flew on the raid, along with Harold 'Micky' Martin, an Australian serving in the RAF. Of these, ten men survived the raid, two (Barlow and Williams) were killed, and one (Burcher) was taken prisoner of war. Left to right: F/L Bob Hay, P/O Lance Howard, F/L David Shannon, F/L Jack Leggo, P/O Frederick Spafford, F/L Harold 'Micky' Martin, P/O Les Knight, F/S Bob Kellow. Not shown in this picture: P/O Bert 'Toby' Foxlee, F/S Thomas 'Tammy' Simpson and P/O Anthony Burcher.

IWM CH9935.

Canadians played a major role in the dams raid. Of the 133 airmen who took part, 29 were Canadian. Thirteen were killed during the raid and one became a prisoner of war. Four of those who survived the mission were killed in action later in the war. F/L Joe McCarthy was an American who flew with the Royal Canadian Air Force. Left to right: Sgt Stefan Oancia, Sgt Fred Sutherland, Sgt Harry O'Brien, F/S Ken Brown, F/S Harvey Weeks, W/O John Thrasher, F/S George Deering, Sgt William Radcliffe, F/S Don MacLean, F/L Joe McCarthy, F/S Grant MacDonald; kneeling: W/O Percy Pigeon, P/O Torger Taerum, F/O Danny Walker, W/O Chester Gowrie, F/O David Rodger.

highlighting this thing, and we were treated like bloody gods. It was daft. I couldn't take it in – neither could anybody else, and there was a lot of drinking going on.

Flying Officer Dave Rodger

Rear gunner, AJ-T

The next day we were pretty numb. It took a while longer than that to realise someone you knew was gone. Some we didn't know well at all because we hadn't been together that long before we went on the trip.

Flight Lieutenant Les Munro

Pilot, AJ-W

We had only been together for six weeks before the raid. It wasn't as if we had been together on a squadron for several months, getting to know people in the mess before and after operations. In this case some of them were quite well known to me, like Henry Maudslay. He was my instructor at 1654 Heavy Conversion Unit at Wigsley. It was sad to know a lot of those blokes, like John Hopgood for instance, a first-class chap, and Rob Barlow the Australian, weren't coming back. There was a sense of sadness, but in time of war, losses were being felt by all the squadrons in Bomber Command operations. You got used to it and you didn't let the loss of individuals that you knew get the better of you and affect your ability to carry out your duty on operations. There was still the feeling of loss when the crews came home and celebrated in the mess, knowing that these other aircrew would never come back.

Flight Sergeant Grant MacDonald

REAR GUNNER, AJ-F

At Scampton we were billeted in what had been station married quarters in the pre-war period. The day after the raid we were pretty stunned when we saw the lorries coming along the houses to pick up the effects of the ones who were gone. Until then we didn't realise it had been so many.

Flying Officer Edward Johnson

BOMB-AIMER, AJ-N

We were shattered at the losses, but somehow we'd expected they'd be fairly high.

Flight Lieutenant David Shannon

PILOT, AJ-L

I suppose we had become hardened to loss – we could shrug it off. We had to, otherwise we could never have flown again. We were debriefed, then we all went back to the Mess and opened the bar and started drinking. The beer started flowing till late in the morning when we struggled off to bed and slept for a few hours. Then there was a stand-down for survivors for a week. The whole thing really took off because it got so much publicity. The press were allowed to write about it, and some brilliant journalist coined the name 'The Dam Busters'.

Barnes Wallis

The brilliant academic and scientist – the peace-loving theorist – became aware, as the crews of eight aircraft did not return, that in practice, his master plan had cost the lives of so many young men. Wrenched by the loss and beyond consolation, the raid changed Barnes Wallis's whole approach to his scientific research in the future.

Flight Sergeant Leonard Sumpter

BOMB-AIMER, AJ-L

I didn't speak to Barnes Wallis after we got back – he was at one end of the room and we were down the other. We sat round the table giving the Intelligence Officer details of the raid and what had happened to us.

Sergeant George 'Johnny' Johnson

BOMB-AIMER, AJ-T

We only had contact with Barnes Wallis on the night before the briefing, when he told us about the bomb, and how it was developed and so on, and again at the briefing when he described what he felt would be required to breach each of the dams.

I don't recall him being in the operations room when we got back, but I'm sure he must have been. I think he stayed until the bitter end to make sure that they were quite certain that none of the crews that were missing would in fact return. He was very emotionally upset by the fact that his invention had caused so much death.

He struck me as being a very determined man – one who was quite convinced of his work and that *it* would work. He was, perhaps, a trifle shy but I suspect this was a result as I read some time later, of his dealings with the hierarchy, which in the early stages were pretty blustery. He had some considerable difficulties in proving his point and getting this particular raid approved.

Flying Officer Edward Johnson

BOMB-AIMER, AJ-N

Wallis was there – but he didn't stay too long because he was very upset at the losses. He was a very sensitive man, Barnes Wallis, and a gentleman, and he couldn't face the fact that we lost so many, once it became known that they had. He went shortly afterwards, with Whitworth – he might have gone for a sleep, and then they sent him home afterwards.

Wallis was a character, but he was very intelligent. He was trying to do something with a real purpose, and I think he'd been a bit frustrated by people not seeing eye to eye with him. He was a genuine, sincere fellow, and everybody got on with him extremely well. All the crews were very fond of him. In appearance he was rather a father figure – a little bit older than most of us – even the older ones. I think you'd describe him as fatherly, rather than looking like a scientist – not flashily dressed, in fact very ordinarily dressed – but obviously a thinking and intelligent man. I think this is why he was so upset at the losses – he was such a thinking and intelligent man.

Flight Lieutenant Les Munro

PILOT, AJ-W

Barnes Wallis – we saw a little bit of him, but not a great deal. He always impressed as a quiet gentleman – very genuine in his beliefs, and the portrayal of him after the raid when he realised how many people had been lost, was quite genuine. He'd never realised what he had taken part in – that the attack on the dams would result in the loss of all those lives. In that respect he was very sincere and I think everybody who came into contact with him felt the influence of his experience.

Sergeant George 'Johnny' Johnson

Bomb-aimer, AJ-T

We had lost eight crews – one hell of a loss for one particular raid and one particular squadron, and in fact, I'm fairly certain that Barnes Wallis actually cried, because he felt he had been the cause of killing so many young aircrew.

Flight Sergeant Ken Brown

Pilot, AJ-F

It was the continuation of an experiment as far as Barnes Wallis was concerned. I don't think he ever expected anyone to get hurt – let alone get killed.

Sergeant George 'Johnny' Johnson

Bomb-aimer, AJ-T

Barnes Wallis actually cried, because he felt he was responsible for the loss of all those young men's lives. He wasn't, of course – but that was the way he felt it. And it was pretty shattering to think we had lost so many, and then to realise how lucky we had been to survive – because we *were* lucky.

Mary Stopes-Roe

Daughter of Barnes Wallis

I don't think he'd envisaged in any clear sense, that outcome – although I supposed he must have known that it would happen. But perhaps the deaths were greater in number than he'd thought, and he was very, very upset. I think it may be that had he known how many young men would go, he might not have set the whole raid in motion.

Flight Lieutenant Les Munro

PILOT, AJ-W

I don't think he really appreciated that we were going to lose anybody. He was concentrating so much on whether his invention would work, and whether his vision of breaching the dams was going to be achieved.

Flying Officer Harold Hobday

NAVIGATOR, AJ-N

Wallis was there at Scampton when we got back, and he was shattered because so many planes were missing. We got used to it, although it was rather more than average – but we were used to planes going missing.

Beck Parsons

GROUND CREW

It shook Barnes Wallis, who didn't expect that sort of loss, but to the ordinary people it was something you got used to, and today was one day and tomorrow was another day. Nobody wanted that sort of loss or expected it, but the various ground crew of the aircraft that were lost, they were badly cut up – but that was yesterday, and we had to move forward. It seems hard but it wasn't really. As soon as anyone went missing, shot down or whatever, they moved all their chairs around in the mess so it wouldn't be so noticeable.

Flight Lieutenant David Shannon

PILOT, AJ-L

Barnes Wallis was a most quiet, unassuming and charming man – one of the most wonderful men and a brilliant, brilliant scientist –

and he was terribly, terribly distressed. He had been sitting in the ops room back at Scampton from the briefing through the day and through the night. With the coded messages we sent, every time we knew an aircraft was gone and we signalled back to base, he heard them. He got the messages when the dams were broken, but although we had proved his bombs would work, he never for a moment thought the cost would be so high. He was in tears and a more distressed figure it would have been hard to imagine by the time the last aircraft had landed. He had not realised that there would be this tremendous sacrifice of life. He was in tears and quite pathetic the following morning.

Barnes Wallis

In a letter responding to Roy Chadwick's congratulatory telegram
I am very deeply grateful, but feel that an enormous share of the credit is due to you. I appreciate all the work which you and your assistants have done, and congratulate you all on the immense success of your efforts. To you, personally, in a special degree, was given the making or breaking of this enterprise . . .

I can assure you that I very nearly had heart failure until you decided to join in the great adventure. No one believed that we should do it. You yourself said it would be a miracle if we did, and I think the whole thing is one of the most amazing examples of teamwork and co-operation in the whole history of the war. May I offer you my very deep thanks for the existence of your wonderful Lancaster, for it was the only aircraft in the world capable of doing the job, and I should like to pay my tribute of congratulation and admiration to you, the designer.

Roy Chadwick

Designer of the Lancaster, in a letter to his daughter

It was a great success, but a great many wonderful young men were killed. Barnes was in tears when I left him.

Barnes Wallis

There is no greater joy in life than first proving that a thing is impossible and then showing how it could be done. Any number of experts had pronounced that the Möhne, and Eder dams could not possibly be destroyed by any known means. And then one shows it can be done – but the doing was done by Guy Gibson and 617 Squadron – not by me.

CHAPTER 12

Celebrations – and Counting the Cost

On regular operations in Bomber Command losses were an everyday occurrence, and the returning crews' celebrations were as much a way of drowning their sense of loss as of their relief in succeeding in their mission and surviving.

Flying Officer Harold Hobday

NAVIGATOR, AJ-N

Once we landed we were debriefed, then we started having drinks – my goodness, it was quite a night. I went to sleep in the mess and woke in the morning in an armchair. After that we were fêted all over the place and had a marvellous time for a week or two, then we went on leave.

Flight Lieutenant Harry Humphries

ADJUTANT, 671 SQUADRON

In the mess they were all round the piano and drinking, and they got hold of me. 'Come on Adj, have a beer.' And of course I did. But I felt bloody awful, really, because I knew a lot of the missing chaps.

Sergeant Frederick Sutherland

FRONT GUNNER, AJ-N

After landing at Scampton, everybody was talking at once, just like a party on a Saturday night. Everybody was excited and we didn't have a chance to shoot our line. Les never shot his line anyway, so we didn't have much of a go at telling all the things we did.

Flight Sergeant Leonard Sumpter

BOMB-AIMER, AJ-L

I think we had a bottle of whisky or brandy that night in the debriefing room – if you wanted whisky in your tea you could have it – which was unusual. Then we sat around and said who we thought we'd seen go down. Someone saw Burpee go down, and someone else saw Astell. There were those we lost at the dams, so we knew we weren't all going to get back. We stayed in the crew room until about five in the morning, then we knew nobody else was going to get back. We just said good morning and went to bed and slept.

When we got up again, two or three hours later we were sitting outside the Sergeants' Mess and chaps from the other squadron came out and asked, 'Have a good time last night, didn't you?' All that joking – then it was all in the newspapers, and the locals knew that the raid had left from Scampton, so it became a famous station. Some of us got awards on that raid – they were immediate. You got a telegram from Arthur Harris in your letter rack. 'You have been awarded an immediate Distinguished Flying Medal' and you'd go round to a shop and buy the little piece of ribbon and put it on your tunic.

Sergeant Ray Grayston

FLIGHT ENGINEER, AJ-N

They had a big drinking do – went on for weeks as far as I know – they visited all the various messes around the country. Les Knight was the taxi pilot most of the time, because he didn't drink, so he flew them out and back.

Clearing up

The ground crews faced the task of repairing the material damage sustained by the returning aircraft, but the task for Gibson and the squadron administration was harder – to them fell the duty of informing fifty-six families of the loss of their loved ones in action.

Kenneth Lucas

GROUND CREW

We only found out what the targets were the day after the raid – Gibson told us – got the whole squadron together. He said as soon as the damaged planes were removed from the 'drome, we could all have three days' special leave – which went down very well.

'Duke' Munro

GROUND CREW FOR MCCARTHY'S AIRCRAFT

We finally discovered one or two aircraft in the north-east corner of the field next to the bomb dump. It was as if these aircraft had managed to land and had their motors cut as soon as they reached the perimeter track. They had flak holes through the fuselage of such a size that you could put your fist through them. The tractors were hitched on and these were hauled back to their hard standing.

Flight Lieutenant Harry Humphries

ADJUTANT, 617 SQUADRON

There were fifty-six families throughout Britain and the Empire who wanted news of their loved ones, and I, unfortunately, had to send that news. I shaved and finished breakfast as quickly as possible. I managed to escape from the mess without any of the boys trying to pour beer down my throat, and my last recollection was of several stubble-chinned, bleary-eyed aircrew types croaking out dirty songs about the Germans around the Mess piano.

I was with the Wingco for some considerable time and gradually we evolved our master-plan for dealing with the amount of work in hand. The first thing was to complete as much as possible on this day, and tomorrow, after the AOC's address, we would follow on with the remainder of the casualty procedure, including all the letters to the next-of-kin.

Accolade and investiture

The success of the daring raid made overnight heroes of the 617 Squadron crews – despite the losses and the failure to break the Sorpe Dam, the mission was a huge boost to public morale and a feat that Churchill, in America to rally support for the war in Europe, was proud to report.

Flight Sergeant George Chalmers

WIRELESS OPERATOR, AJ-O

The next thing we knew, the King and Queen were coming up to the camp and there was all this calling us the 'glamour boys'. They came and went, and then the next thing we were going down for investiture to collect the medals. It was one thing after another and you couldn't really believe it. Even to this day. The British people lapped it up and thought it was marvellous.

Doug Godfrey

Fitter, Avro

After the raid the news flashed round the factory. Cheers rang out, and shortly afterwards our chairman, Roy Dobson, came round to add his congratulations. Our hearts swelled with pride. Then Guy Gibson and his crew came and thanked us – something I will never forget.

Sergeant Dudley Heal

Navigator, AJ-F

After the raid, we felt we'd done the best we could, and achieved a certain amount of success. Two dams had been breached, though ours hadn't. We were disappointed about that. As a matter of fact, I felt that we might even be expected to go and have another go at it. So, when we were sent on a week's leave immediately after the raid and I was sitting with my family at tea on the Wednesday, the doorbell rang and I went to see a telegraph boy with a telegram for me from Scampton. I walked back into the living room, showed my family the telegram and said, 'It rather looks as though I've got to go back.' And I opened the telegram and it said, 'Heartiest congratulations on the award of the Distinguished Flying Medal. Wingco.' And my family and I looked at each other and it was impossible to say who was the most surprised and relieved.

Sergeant Jim Heveron

In charge of the orderly room

Several of us received Mentions in Despatches when the aircrew were decorated, but in my opinion the whole ground staff should have been recognised.

Air Vice-Marshal the Hon R A Cochrane

Message to Gibson:

All ranks in 5 Group join me in congratulating you and all in 617 Squadron on a brilliantly conducted operation. The disaster which you have inflicted on the German war machine was a result of hard work, discipline and courage. The determination not to be beaten in the task and getting the bombs exactly on the aiming point in spite of opposition have set an example others will be proud to follow.

Cochrane also wrote to Barnes Wallis:

Before reaching the end of this somewhat long but exciting day, I felt I must write to tell you how much I admire the perseverance which brought you the astounding success which was achieved last night. Without your determination to ensure that a method which you knew to be technically sound was given a fair trial, we should not have been able to deliver the blow which struck Germany last night.

Sir Henry Tizard

Chief of the Air Staff

In a letter to Barnes Wallis:

Taking it all in all, from the first brilliant ideas, through the model experiments and the full-scale trials, remembering also that when the sceptics were finally convinced you had to work at the highest pressure to get things done in time, I have no hesitation in saying that yours is the finest technical achievement of the war.

On 27 May 1943, HM King George VI visited Scampton where, with, left to right, Cochrane, Gibson, Group Captain Whitworth and another officer, he was shown the model of the Möhne Dam.

Courtesy of www.ww2images.com.

Guy Gibson and members of 617 Squadron with the designer of the Lancaster, Roy Chadwick.

Mary Stopes-Roe

BARNES WALLIS'S DAUGHTER

News of the dams raid came to me as a surprise. The house-mistress at my boarding school said cheerfully, 'So your father has done it again.' The year before, my father had given a lecture at the school on the problem of high-altitude flying – a lecture so absorbing that my mother had to remind him of the time. The press reported the fact and the outcome of the raid, but not, of course, any detail of the secret weapon used. I was nonplussed, but put two and two together and sent him a telegram, 'Wonderful Daddy'.

Throughout, my parents supported and cherished each other. For myself, they succeeded in protecting us from fear, anxiety, hunger or distress. I admire their generation, and am grateful to have known in small part what they achieved.

Barnes Wallis

In a letter to Cochrane:
It is impossible to find words adequately to express what one feels about the aircrews. The gallantry with which they go into action is incomparable. While the older generation of Air Force officers may not be called upon to carry out actual attacks in person, the spirit of their juniors must proceed from their thought and training, and in praising your crews, I would like to add the thanks which I feel are due to you as one of the senior officers of the Air Force, for the outstanding generation of pilots which your example and training has produced. Will you please accept the deepest sympathy of all of us on the losses which the Squadron has sustained. You will understand, I think, the tremendous strain which I have felt at having been the cause of sending these crews on so perilous a mission – and the tense moments in the Operation Room when,

after four attacks, I felt that I had failed to make good, were almost more than I could bear; and for me the subsequent success was almost completely blotted out by the sense of loss of those wonderful young lives. In the light of our subsequent knowledge, I do hope that all those concerned will feel that the results achieved have not rendered their sacrifice in vain.

Flight Lieutenant Harry Humphries

ADJUTANT, 617 SQUADRON

The main event came on 22 June, when all those awarded for their part in the 'Dam Raid' received news of their summons to Buckingham Palace. This was a great event and it was decided to organise a mass visit to the metropolis for the occasion. I was asked to organise a special train. Our friends, the RTOs, co-operated very well and fixed us up without much trouble. The station Motor Transport Officer promised a special bus for the journey to and from the LNER station. The manufacturers of the Lancaster aircraft, Merssrs A V Roe, put the finishing touches to our plans by promising a special party in London in the evening following the investiture.

On the train the screw-tops began to come off beer bottles and the party started. When we arrived at Grantham, strangely enough we still had a full contingent. Usually on occasions like this someone tries to jump out of a carriage window for a bet. We did have quite a bit of fun on the station. Our carriages had to be shunted on to the London train. Jack Fort and one or two others decided they would like to drive the train. We eventually found them streaked in grime, but very happy, in the driver's cabin.

This time it was not so easy to get them together. At train departure time, some were in the buffet, some in the toilet – and others had just disappeared. When the train drew out I had a quick check and much to my relief we were still one hundred per cent present.

I decided that from now on I would let matters take their own course, otherwise I would finish up a nervous wreck.

John Elliott

NCO GROUND CREW

After the raid there was a lot of publicity. The first thing we had after it was an investiture to receive all the gongs that they got and on the evening of that investiture we had a party in Regent Street for the aircrew and I think there were two of us, two senior NCOs from each flight on the squadron.

Sergeant Ray Grayston

FLIGHT ENGINEER, AJ-N

We were successful on the raid and didn't suffer any damage at all, and they awarded Les Knight a DSO, and two other members of the crew, the navigator and the bomb-aimer both got the DFC, and the rest of us got nothing, as per usual. They can't fly without us, but they didn't recognise these odd bods in the crew.

Flight Sergeant Leonard Sumpter

BOMB-AIMER, AJ-L

One thing that got up my nose after the raid was when the King and Queen came down to Scampton, and all the photographs were taken with the officers in front of the Officers' Mess. But the flight sergeants and sergeants didn't see a sign of the King and Queen – they didn't come near our mess. Yet there were more NCOs on the raid than officers. And I thought this was a bit much. I didn't really mind, but I thought they might have come over and said hello to us.

Courtesy of www.ww2images.com.

Gibson and his wife Eve celebrate with fellow officers on their visit to London.

Flight Sergeant George Chalmers

WIRELESS OPERATOR, AJ-O

Naturally, after the raid, there was a certain sadness about it all. But we were sent on leave almost immediately. We were back in about a week, to go to the Palace to get decorated. It was then, I suppose, that amongst all the celebration we realised that certain chaps were not there that should be there. Even while we were living it up a bit, I think there was a feeling of sadness.

Gwyn Johnson

WIFE OF JOHNNIE JOHNSON

On the way back from the London investiture, the whole squadron were on this train and absolutely bladdered – and none of them had any money, so I paid for them all to have a cup of tea and a bun on the way back from London to Lincoln.

Retrospective

When the immediate blaze of publicity died down, there was time to consider the raid – and weigh the achievement against the losses and the extent to which German industry had been crippled.

Flight Lieutenant Harold 'Micky' Martin

PILOT, AJ-P

The effect of the dams raid on the public morale was tremendous. It seemed to catch their whole imagination. And it was proof, too, of a cohesion, so necessary in war, where you have to tie together so many threads of human activity. It starred the brilliant scientist Barnes Wallis and his inventions and coupled with them the exceptional organising power and drive of Sir Arthur Harris and the brilliant leadership of a young man, Guy Gibson.

And this showed the British public that from every quarter and every section they could get together for really clouting the Nazis and winning the war.

Group Captain Douglas Bader

In German POW camp at the time of the raid

I well remember the destruction of the Möhne and Eder dams while I was in a prison camp. It had an enormous effect on the Germans and the opposite effect, of course, on us, the prisoners of war.

Flight Sergeant Leonard Sumpter

Bomb-aimer, AJ-L

We learned after the raid that the Germans increased the armament on the dams in case we went again to try to damage them while they were repairing them. I could never understand why we never went again, or ever tried high-level bombing – but nobody ever bothered about that, and the dams were back in operation about six months later. Whereas, if they had made nuisance raids and gone over about once a week with about a hundred decent bombers the repairs could have been prevented. I could never understand why they didn't do this.

Flight Lieutenant David Shannon

Pilot, AJ-L

In hindsight, I think the training was absolutely adequate for the job. The limiting factors were the time – one can go on and on and on, but we'd probably averaged a hundred hours per crew training – specific training with the formation flying, low-level flying, first at 150 feet, then down to 60 feet. We had all dropped mines – all had practice on VHF radio communications. I don't think any further training would have achieved any different result.

It's easy to say, after the event, but the Germans were able to rebuild the dams with much more speed than had been thought by the pundits back home – and it turns out that the reason this was possible was that the explosions and the water took away the dam walls, but left the foundations of the dams untouched – so all they had to do was rebuild on the existing foundations. Perhaps if Bomber Command had gone back the following night or within a very short space of time – and the Germans themselves have said the same thing, and couldn't understand why it wasn't done – that if they'd blown away the foundations with deep-penetration bombs when the water had gone, there was no way they could have been rebuilt.

George Robert Edwards

VICKERS AIRCRAFT ENGINEER

When the aircraft came back, it was fairly obvious that it had worked and we'd actually blown some holes in these dams. The extent to which those dams were part of the vital war effort of the Germans has always been argued. A lot of people maintained that they didn't really count – but the object of the exercise was to blow a hole in the dams – and that it did. Subsequent value to the war effort, and how many people in the Ruhr district got drowned, was something you didn't think about at the time. It was the clinical approach where you set out to blow a hole in the dam wall and if you did that, you reckoned you'd done all right.

The dam buster raids had an enormous morale-boosting effect. Things weren't going very well and this looked as if we were capable of giving the enemy a punch on the nose, and that was pretty important.

Flight Sergeant George Chalmers

Wireless operator, AJ-O

Looking back on the raid, it's difficult to put it into any words that would make any sense. Nothing that happened in the war made sense to my way of thinking. I suppose people who knew better than me have already put it more accurately than I could, and that is to say that it caused a great deal of trouble for the Germans – which is what we intended to do. It probably affected the length of the war to some extent. It certainly drew an awful lot of people away from the war front to clean up the mess, for one thing. It also slowed down a lot of armament factories as well. It didn't stop them – but it certainly slowed them down. So in that sense, I think it was worthwhile, but it's difficult to assess people's lives.

John Elliott

NCO ground crew

I suppose the losses – if losses can ever be worth it – were worth it. I don't think there're any two ways about it. It was probably the single most destructive raid carried out by a squadron during the war. They did some after that but not anything as destructive as that.

When it all calmed down we were quite pleased about the whole thing – proud of the squadron. We had to take a bit of flak later on from some of the other squadrons because some time after that we got labelled as being a one-operation squadron. That didn't go down very well but we didn't take a lot of notice of it really.

Flight Lieutenant Les Munro

PILOT, AJ-W

On reflection afterwards I felt a sense of relief that because I was hit and had to return to base, it probably saved my life. Even to this day I think if I had gone on there was a strong possibility I wouldn't have come back. If you look at the percentage of the ones who got there and didn't come back, the possibility is that if I hadn't been hit by flak, I would have been among those who didn't come back.

After the mission, there have been different opinions expressed that the operation itself was not successful, and that the ends did not justify the means, with the loss of eight crews. But I refute that – I say the raid was very successful in the operational sense. What you could say is that perhaps the War Ministry, in deciding that the dams should be attacked and breached did not appreciate that the Germans would be capable of repairing the damage to the dams in so short a time as they did – something like four or five months, and in that respect maybe it wasn't a complete success, but I've no doubt that it caused complete devastation. Factories were flooded, the infrastructure was taken out, roads and bridges and so on, and while that might have been repaired over a period of months, there was still a lot of damage caused with a fair loss of life. Unfortunately forced labour was part of the human loss, but I believe that the greatest value was for the morale of the British people as a result of the operation. I think it was successful in that sense.

And from the operational point of view, it was also successful, in spite of the losses. We had the primary targets, the Möhne and Eder, both breached but the Sorpe was not damaged to the extent we hoped it would be. Joe McCarthy and the Canadian Ken Brown both bombed the Sorpe, and while Wallis had believed that if two or three Upkeeps could be rolled down the bank of the Sorpe and exploded at a depth of 30 feet of water, the earthquake effect might crack the concrete wall – that might be sufficient to start seepage and cause ongoing problems.

In hindsight I was very pleased that I flew Lancasters and not Halifaxes or the Sterling. I have a great deal of time for the Lancaster – it was a wonderful aircraft and had no vices. It was easy to fly, and it was capable of carrying large bombloads and flying vast distances, and I give it full marks. I still believe it was the best heavy bomber used in the war.

John Elliott

NCO GROUND CREW

After the dams raid we moved from Scampton to Coningsby. Gibson had gone and we did two more really abortive raids on the Dortmund Ems Canal. The first effort was aborted because of the weather and we lost our Flight Commander, David Maltby – he went into the sea. And the second time they went the weather was believed to have been good but apparently over the target area it was bad and instead of coming back they pressed on with it and the losses were pretty colossal. It was a shocking loss to have achieved nothing. Nothing was achieved by either of the raids and we lost a lot of really top-notch crews.

Roy Chadwick

DESIGNER OF THE LANCASTER

I shall always remember this particular operation as an example of how the engineers of this country have contributed substantially towards the defeat of our enemies.

A group at a dams raid dinner given by A V Roe and Co.

Key: 1. Sir Charles Craven, Chairman, Vickers-Armstrong; 2. Sir Hew Kilner, Works Manager, Vickers-Armstrong, 3. Air Cdre Charles Whitworth, OC RAF Scampton; 4.F/L Les Munro (AJ-W); 5. Capt H A 'Sam' Brown, Test Pilot, Avro; 6. F/L 'Capable' Caple, 617 Squadron engineering officer; 7. F/S Heveron; 8. F/L Harold Wilson; 9. F/S 'Chiefy' Powell, 10. F/L Richard Trevor-Roper (AJ-G); 11. F/O Edward Johnson (AJ-N); 12. F/O Harold Hobday (AJ-N); 13. F/L Harry Humphries, 617 Squadron adjutant; 14. P/O Don MacLean (AJ-T); 15. Capt Joe 'Mutt' Summers, Chief Test Pilot, Vickers-Armstrong; 16. Barnes Wallis; 17. Sir Roy Dobson, Managing Director, Avro; 18. P/O Lance Howard (AJ-O); 19. F/L Mick Martin (AJ-P); 20. F/L Jack Leggo (AJ-P); 21. Sgt Vivian Nicholson (AJ-J); 22. F/L David Maltby (AJ-J); 23, F/L Joe McCarthy (AJ-T) 24. W/C Guy Gibson (AJ-G); 25. F/S Len Sumpter (AJ-L); 26. P/O Toby Foxlee (AJ-P); 27. F/L Bob Hay (AJ-P); 28. F/S Tammy Simpson (AJ-P); 29. Roy Chadwick, Chief Designer, Avro; 30. Sir Frank Spriggs, Chairman, Glosters; 31. Sir Thomas Sopwith, Chairman, Hawker Siddeley.

Flying Officer Harold Hobday

NAVIGATOR, AJ-N

After all Barnes Wallis's efforts, we didn't dream the thing would fail, and of course, it didn't. There were several bombs which were launched which didn't actually hit the target, or that didn't break the target – but that was because they were off the right position. They didn't burst in the right place. The ones that burst in the right place broke the walls of the two dams straight away. He proved that afterwards. I talked to him after the war and he'd been out there and looked at where all the bombs had burst. He said that both the Eder and the Möhne dams were broken by one bomb each.

Looking back, there have been so many different ideas about the raid. I think it was a good thing, especially from a morale point of view. I understand that if the Sorpe Dam, which was the third one to be attacked, had been broken, the effect on the steel industry would have been much worse than it was in Germany – but I still think it was well worth doing. When you consider the atmosphere during the war, and how morale could be boosted – or could drop so easily – it was a vital thing to keep morale up. The dams raid, apart from anything else, did a lot for the morale of the country and abroad. It also did a lot of damage in Germany – I'm sure it did. So when people say, 'Was it worth it?' I say that it was.

APPENDIX I

The Crews

The First Wave

AJ-G 'George'

Pilot	Wing Commander Guy Gibson, DSO and Bar, DFC and Bar
Navigator	Pilot Officer Torger Taerum (RCAF)
Bomb-aimer	Frederick Spafford, DFM (RAAF)
Flight engineer	Sergeant John Pulford
Wireless operator	Flight Lieutenant Robert Hutchison, DFC
Front gunner	Flight Sergeant Andrew Deering (RCAF)
Rear gunner	Flight Lieutenant Richard Trevor-Roper, DFM

AJ-M 'Mother'

Pilot	Flight Lieutenant John 'Hoppy' Hopgood, DFC and Bar
Navigator	Flying Officer Ken Earnshaw (RCAF)
Bomb-aimer	Flight Sergeant John Fraser (RCAF)

Flight engineer	Sergeant Charles Brennan
Wireless operator	Sergeant John Minchin
Front gunner	Pilot Officer George Gregory, DFM
Rear gunner	Pilot Officer Anthony Burcher, DFM (RAAF)

AJ-P 'Popsie'

Pilot	Flight Lieutenant Harold 'Micky' Martin, DFC
Navigator	Flight Lieutenant Frederick Leggo, DFC (RAAF)
Bomb-aimer	Flight Lieutenant Robert Hay, DFC (RAAF)
Flight engineer	Pilot Officer Ivan Whittaker
Wireless operator	Flying Officer Leonard Chambers (RNZAF)
Front gunner	Pilot Officer Bertie Foxlee, DFM (RAAF)
Rear gunner	Flight Sergeant Thomas Simpson (RAAF)

AJ-A 'Apple'

Pilot	Squadron Leader Melvin 'Dinghy' Young, DFC and bar
Navigator	Flight Sergeant Charles Roberts
Bomb-aimer	Flying Officer Vincent MacCausland (RCAF)
Flight engineer	Sergeant David Horsfall
Wireless operator	Sergeant Lawrence Nichols
Front gunner	Sergeant Gordon Yeo
Rear gunner	Sergeant Wilfred Ibbotson

AJ-J 'Johnny'

Pilot	Flight Lieutenant David Maltby, DFC
Navigator	Sergeant Vivian Nicholson
Bomb-aimer	Pilot Officer John Fort
Flight engineer	Sergeant William Hatton

Wireless operator	Sergeant Anthony Stone
Front gunner	Sergeant Victor Hill
Rear gunner	Sergeant Harold Simmonds

AJ-Z 'Zebra'

Pilot	Squadron Leader Henry Maudslay, DFC
Navigator	Flying Officer Robert Urquhart, DFC (RCAF)
Bomb-aimer	Pilot Officer Michael Fuller
Flight engineer	Sergeant John Marriott, DFM
Wireless operator	Warrant Officer Alden Cottam
Front gunner	Flying Officer William Tytherleigh, DFC
Rear gunner	Sergeant Norman Burrows

AJ-L 'Leather'

Pilot	Flight Lieutenant David Shannon, DFC (RAAF)
Navigator	Flying Officer Daniel Walker, DFC (RCAF)
Bomb-aimer	Flight Sergeant Leonard Sumpter
Flight engineer	Sergeant Robert Henderson
Wireless operator	Flying Officer Brian Goodale, DFC
Front gunner	Sergeant Brian Jagger
Rear gunner	Flying Officer Jack Buckley

AJ-B 'Baker'

Pilot	Flight Lieutenant William Astell, DFC
Navigator	Pilot Officer Floyd Wile (RCAF)
Bomb-aimer	Flying Officer Donald Hopkinson
Flight engineer	Sergeant John Kinnear
Wireless operator	Warrant Officer Abram Garshowitz (RCAF)
Front gunner	Flight Sergeant Francis Garbas (RCAF)
Rear gunner	Sergeant Richard Bolitho

AJ-N 'Nuts'

Pilot	Pilot Officer Leslie Knight (RAAF)
Navigator	Flying Officer Harold Hobday
Bomb-aimer	Flying Officer Edward Johnson
Flight engineer	Sergeant Raymond Grayston
Wireless operator	Flight Sergeant Robert Kellow (RAAF)
Front gunner	Sergeant Frederick Sutherland
Rear gunner	Sergeant Harry O'Brien

AJ-E 'Easy'

Pilot	Flight Lieutenant Robert Barlow DFC (RAAF)
Navigator	Flying Officer Phillip Burgess
Bomb-aimer	Pilot Officer Alan Gillespie DFM
Flight engineer	Pilot Officer Samuel Whillis
Wireless operator	Flying Officer Charles Williams, DFC (RAAF)
Front gunner	Flying Officer Harvey Glinz (RCAF)
Rear gunner	Sergeant Jack Liddell

The Second Wave

AJ-T 'Tommy'

Pilot	Flight Lieutenant Joseph McCarthy, DFC (RCAF)
Navigator	Flying Officer Donald MacLean (RCAF)
Bomb-aimer	Sergeant George 'Johnny' Johnston
Flight engineer	Sergeant William Radcliffe (RCAF)
Wireless operator	Flight Sergeant Leonard Eaton
Front gunner	Sergeant Ronald Batson
Rear gunner	Flying Officer David Rodger (RCAF)

AJ-W 'Willie'

Pilot	Flight Lieutenant John Leslie Munro (RNZAF)
Navigator	Flying Officer Grant Rumbles
Bomb-aimer	Sergeant James Clay
Flight engineer	Sergeant Frank Appleby
Wireless operator	Flying Officer Percy Pigeon (RCAF)
Front gunner	Sergeant William Howarth
Rear gunner	Flight Sergeant Harvey Weeks (RCAF)

AJ-K 'King'

Pilot	Pilot Officer Vernon Byers (RCAF)
Navigator	Flying Officer James Warner
Bomb-aimer	Pilot Officer Arthur Whittaker
Flight engineer	Sergeant Alistair Taylor
Wireless operator	Sergeant John Wilkinson
Front gunner	Sergeant Charles Jarvie
Rear gunner	Flight Sergeant James McDowell (RCAF)

AJ-H 'Harry'

Pilot	Pilot Officer Geoffrey Rice
Navigator	Flying Officer Richard MacFarlane
Bomb-aimer	Warrant Officer John Thrasher (RCAF)
Flight engineer	Sergeant Edward Smith
Wireless operator	Warrant Officer Chester Gowrie (RCAF)
Front gunner	Sergeant Thomas Maynard
Rear gunner	Sergeant Stephen Burns

The Third Wave

AJ-C 'Charlie'

Pilot	Pilot Officer Warner Ottley, DFC

Navigator	Flying Officer Jack Barrett, DFC
Bomb-aimer	Flight Sergeant Thomas Johnston
Flight engineer	Sergeant Ronald Marsden, DFM
Wireless operator	Sergeant Jack Guterman, DFM
Front gunner	Sergeant Harry Strange
Rear gunner	Sergeant Frank Tees

AJ-S 'Sugar'

Pilot	Pilot Officer Lewis Burpee, DFM (RCAF)
Navigator	Sergeant Thomas Jaye
Bomb-aimer	Flight Sergeant James Arthur (RCAF)
Flight engineer	Sergeant Guy Pegler
Wireless operator	Pilot Officer Leonard Weller
Front gunner	Sergeant William Long
Rear gunner	Warrant Officer Joseph Brady (RCAF)

AJ-F 'Freddie'

Pilot	Flight Sergeant Kenneth Brown (RCAF)
Navigator	Sergeant Dudley Heal
Bomb-aimer	Sergeant Stefan Oancia (RCAF)
Flight engineer	Sergeant Harry Basil Feneron
Wireless operator	Sergeant Harry Hewstone
Front gunner	Sergeant Daniel Allatson
Rear gunner	Flight Sergeant Grant McDonald (RCAF)

AJ-O 'Orange'

Pilot	Flight Sergeant William Townsend DFM
Navigator	Pilot Officer Lancel Howard RAAF
Bomb-aimer	Sergeant Charles Franklin DFM
Flight engineer	Sergeant Dennis Powell
Wireless operator	Flight Sergeant George Chalmers
Front gunner	Sergeant Douglas Webb
Rear gunner	Sergeant Raymond Wilkinson

AJ-Y 'York'

Pilot	Flight Sergeant Cyril Anderson
Navigator	Sergeant John Nugent
Bomb-aimer	Sergeant John Green
Flight engineer	Sergeant Robert Patterson
Wireless operator	Sergeant William Bickle
Front gunner	Sergeant Eric Ewan
Rear gunner	Sergeant Arthur Buck

Honours and Awards

Thirty-four of the aircrew who took part in the raid were decorated for their actions.

Victoria Cross	Guy Gibson
Distinguished Service Order	Martin, Shannon, Maltby, McCarthy, Knight
Bar to Distinguished Flying Cross	Walker, Hutchison, Leggo, Hay
Distinguished Flying Cross	Chambers, Howard, Deering, Taerum, Spafford, Trevor-Roper, Fort, Hobday, E Johnson, Buckley
Conspicuous Gallantry Medal	Townsend, Brown
Bar to Distinguished Flying Medal	Franklin
Distinguished Flying Medal	G Johnson, Chalmers, Nicholson, Wilkinson, Pulford, Webb, Sumpter, Simpson, Heal, MacLean, Oancia

The Royal Air Force commanders

Air Chief Marshal Sir Charles Portal, Chief of the Air Staff

Air Marshal Sir Arthur Harris, Air Officer Commander-in-Chief, Bomber Command

Air Vice-Marshal R H M S Saundby, Deputy Air Officer Commanding-in-Chief, Bomber Command

Air Vice-Marshal the Honorable Ralph Cochrane, Air Officer Commanding, 5 Group, Bomber Command

Group Captain J N H Whitworth, Station Commander, RAF Scampton

Scientists and experts

Captain J 'Mutt' Summers, test pilot for Vickers-Armstrong

Sir Henry Tizard, scientific adviser to the Ministry of Aircraft Production

Lord Cherwell (formerly Professor F A Lindemann), scientific adviser to Churchill

Dr W H Glanville, Director of Road Research Laboratory (RRL), Harmondsworth

A R Collins, scientific officer in Concrete Section, RRL

Sir Charles Craven, head of Vickers-Armstrong, for whom Wallis worked

Roy Chadwick, aircraft designer for Avro – designed the Lancaster

Barnes Wallis, chief engineer of Vickers-Armstrong

Appendix II

The Fate of the Dambusters

Fifty-three men lost their lives on the night of the raid, as eight Lancasters failed to return to Scampton. Three crewmen managed to bale out: Flight Sergeant John Fraser and Pilot Officer Anthony Burcher from Flight Lieutenant Hopgood's crew and Sergeant Fred Tees from Pilot Officer Ottley's crew – all were taken prisoner of war.

Those killed in action

AJ-B 'Baker'
Flight Lieutenant William Astell
Pilot Officer Floyd Wile
Flying Officer Donald Hopkinson
Sergeant John Kinnear
Warrant Officer Abram Garshowitz
Flight Sergeant Francis Garbas
Sergeant Richard Bolitho

AJ-M 'Mother'
Flight Lieutenant John 'Hoppy' Hopgood
Flying Officer Ken Earnshaw
Sergeant Charles Brennan
Sergeant John Minchin
Pilot Officer George Gregory

AJ-Z 'Zebra'
Squadron Leader Henry Maudslay
Flying Officer Robert Urquhart
Pilot Officer Michael Fuller
Sergeant John Marriott
Warrant Officer Allen Cottam
Flying Officer William Tytherleigh
Sergeant Norman Burrows

AJ-A 'Apple'
Squadron Leader Melvin 'Dinghy' Young
Flight Sergeant Charles Roberts
Flying Officer Vincent MacCausland
Sergeant David Horsfall
Sergeant Lawrence Nichols
Sergeant Gordon Yeo
Sergeant Wilfred Ibbotson

AJ-K 'King'
Pilot Officer Vernon Byers
Flying Officer James Warner
Pilot Officer Arthur Whittaker
Sergeant Alistair Taylor
Sergeant John Wilkinson
Sergeant Charles Jarvie
Flight Sergeant James McDowell

AJ-E 'Easy'
Flight Lieutenant Robert Barlow
Flying Officer Phillip Burgess
Pilot Officer Alan Gillespie
Pilot Officer Samuel Whillis
Flying Officer Charles Williams
Flying Officer Harvey Glinz
Sergeant Jack Liddell

AJ-S 'Sugar'
Pilot Officer Lewis Burpee DFM
Sergeant Thomas Jaye
Flight Sergeant James Arthur
Sergeant Guy Pegler
Pilot Officer Leonard Weller
Sergeant William Long
Warrant Officer Joseph Brady

AJ-C 'Charlie'
Pilot Officer Warner Ottley
Flying Officer Jack Barrett
Flight Sergeant Thomas Johnston
Sergeant Ronald Marsden
Sergeant Jack Guterman
Sergeant Harry Strange

Those killed later in action

23–24 SEPTEMBER 1943: RAID ON MANNHEIM

AJ-Y
Flight Sergeant Cyril Anderson

Sergeant John Nugent
Sergeant John Green
Sergeant Robert Patterson
Sergeant William Bickle
Sergeant Eric Ewan
Sergeant Arthur Buck

14–15 SEPTEMBER 1943: RAID ON DORTMUND-EMS CANAL

AJ-J
Flight Lieutenant David Maltby
Sergeant Vivian Nicholson
Pilot Officer John Fort
Sergeant William Hatton
Sergeant Anthony Stone
Sergeant Victor Hill
Sergeant Harold Simmonds

15–16 SEPTEMBER 1943: RAID ON DORTMUND–EMS CANAL

AJ-G
Piloted by Squadron Leader George Holden
Pilot Officer Torger Taerum
Pilot Officer Frederick Spafford
Flight Lieutenant Robert Hutchinson
Flight Sergeant Andrew Deering

AJ-N
Pilot Officer Leslie Knight
(Knight stayed with his aircraft while his crew escaped, and tried to make a forced landing, but crashed and was found dead at the controls. Locals buried him in the Den Ham Cemetery in Holland.

His crew baled out safely – Grayston and O'Brien were taken prisoner of war, but Hobday, Johnson, Kellow and Sutherland evaded capture and returned to the UK.)

20–21 DECEMBER 1943: RAID ON LIEGE

AJ-H
(Pilot Officer Geoffrey Rice baled out and was taken prisoner of war – the rest of the crew perished)
Flying Officer Richard MacFarlane
Warrant Officer John Thrasher (RCAF)
Sergeant Edward Smith
Warrant Officer Chester Gowrie (RCAF)
Sergeant Thomas Maynard
Sergeant Stephen Burns

1944

12–13 FEBRUARY 1944: ATTACK ON ANTHEOR VIADUCT

AJ-P
Flight Lieutenant Robert Hay fatally shot in aircraft

13 FEBRUARY 1944

Sergeant John Pulford killed in an accident

30 APRIL 1944

AJ-L
Sergeant Brian Jagger killed in flying accident

30–31 March 1944: raid on Nuremberg

AJ-G
Flight Lieutenant Richard Trevor-Roper

19–20 September 1944: attack on Monchengladbach and Rheydt

AJ-G
Wing Commander Guy Gibson

ACKNOWLEDGEMENTS

My initial thanks are to Ed Faulkner, Editorial Director of Virgin Books, who commissioned me to write this oral history of the Dambusters, and to Davina Russell for her skills in editing the text, and Sophia Brown for her excellent work on the photographs.

Throughout the writing of this book I have had the remarkable assistance of Vicky Thomas, whose energy, enthusiasm and consummate skill – not to mention her German translation abilities – have greatly enhanced this work. I would also like to profoundly thank Robert Owen, Official Historian, No 617 Squadron Aircrew Association, who very kindly read through the manuscript and made a number of changes and suggestions, all of which I was happy to incorporate. He has also made it possible for me to contact the five surviving Dambusters. I am greatly in his debt. I am also particularly grateful to the distinguished military historian, John Sweetman, who gave me permission to use extracts from his landmark history, *The Dambusters Raid*, which also furnished a detailed framework and chronology.

I would also like to thank Bruce Hunter and Georgia Glover at

David Higham Associates Ltd, for their kind permission to use extracts from Guy Gibson's outstanding book, *Enemy Coast Ahead*; Kevin Wilson, who has allowed me to use excerpts from his very fine book, *Bomber Boys*; Toby Groom and Jo Mitchell at the History Channel, for use of their documentary interviews; the ever-helpful Sound Archive of the Imperial War Museum; the archives of the RAF Museum; Simon Braithwaite at the Second World War Experience Centre; and the Lancaster Museum, Alberta, Canada. I would particularly like to thank the following publishers for permission to reproduce extracts: *Above All Unseen*, Edward Leaf, Haynes Publishing; *Barnes Wallis: Dambuster*; Peter Pugh, Icon Books; *Dambusters – the Definitive History of 617 Squadron at War, 1943–1945*, Chris Ward, Andy Lee and Andreas Wachtel, Red Kite; *Breaking the German Dams: Flying into History*, Richard Morris and Robert Owen, Newsdesk Communications Ltd; *Living with Heroes*, Harry Humphries, The Erskine Press; *The Dambusters*, John Sweetman with David Coward and Gary Johnstone, Time Warner Books; *The Dam Busters*, Paul Brickhill, Pan Books; *The Dambusters at 60*, Lincolnshire Echo; Tim Robinson, editor of the *Grantham Journal* for passages from *The Dambusters 50 Years*, souvenir issue; *We Will Remember Them*, Jan van den Driesschen with Eve Gibson, The Erskine Press; *Als Deutschlands Dämme brachen*, Helmut Euler, Motorbuch Verlag, Stuttgart.

My thanks also to Channel 4, Windfall Films, the *Daily Mirror* and Delta Music for access to recorded interview material from their respective Dambusters videos/DVDs.

My gratitude also to my dear friends Sir Martin and Lady Gilbert for their support, and for the use of the map of the raid, taken from his newly published title, *The Routledge Atlas of the Second World War*, as drawn by Neil Hanson. A special thank-you, too, to Julia Pickles who transcribed many hours of recorded interviews.

Throughout the writing of this book, my very dear friend Ruth Cowen has, as always, given me great support, as have Don and

Liz McClen and Susan Jeffreys – and Lucia Corti has given me much love and laughter.

Finally, I owe a profound debt of thanks to the five surviving men of 617 Squadron who took part in the raid: Les Munro, George 'Johnny' Johnson, Ray Grayston, Grant McDonald and Fred Sutherland, who have kindly allowed me to use their stories. They are the last witnesses to an extraordinary and daring operation, inspiring in both its conception and execution, which will live for ever in the annals of our nation's history.

Glossary

AAD Committee	Air Attack on Dams Committee
Ack-ack	Anti-aircraft fire
AOC	Air Officer Commanding
CO	Commanding Officer
DFC	Distinguished Flying Cross
DFM	Distinguished Flying Medal
DSO	Distinguished Service Order
Elsan	Chemical toilet
Formate/formating on	Flying in formation with reference to other aircraft
Flying 'O' Badge	Earned after qualifying as an RAF Observer
Highball	The smaller version of Barnes Wallis' Upkeep bomb, designed for use against German shipping
LNER	London North Eastern Railway
MAP	Ministry of Aircraft Production
MT	Motor Transport

NPL	National Physical Laboratory
R/T	Radio telegraphy / telephony
RTO	Railway Transport Officer
SASO	Senior Air Staff Officer
Vic	A 'V' shaped formation of aircraft

Bibliography

Brickhill, Paul, *The Dam Busters*, Pan Books, London, 1999

van den Driesschen, Jan with Eve Gibson *We Will Remember Them*, The Erskine Press, Norwich, 1979

Euler, Helmut, *Als Deutschlands Dämme Brachen*, Motorbuch Verlag, Stuttgart,1982

Gibson, Guy, VC, DSO, DFC, *Enemy Coast Ahead*, Crécy Publishing Limited, Manchester, 2003

Harris, Air Marshal Sir Arthur, *Bomber Offensive*, Pen & Sword Military Classics, Barnsley, South Yorkshire, 2005

Humphries, Harry, *Living with Heroes*, Erskine Press, Norwich, 2003

Leaf, Edward, *Above All Unseen*, Patrick Stephens Limited, Sparkford, Nr Yeovil, Somerset, 1997

Pugh, Peter, *Barnes Wallis: Dambuster*, Icon Books, Cambridge, 2005

Sweetman, John, *The Dambusters Raid*, Cassell, London, 1999

Sweetman, John with David Coward and Gary Johnstone, *The Dambusters*, Time Warner Books, London, 2003

Ward, Chris, Andy Lee and Andreas Wachtel, *Dambusters – the Definitive History of 617 Squadron at War, 1943–1945*, Red Kite, Walton on Thames, 2003

Wilson, Kevin, *Bomber Boys*, Orion Books, London, 2005

The Dambusters 50 Years, *Grantham Journal* souvenir issue, 1993

The Dambusters at 60, *Lincolnshire Echo*, 8 April 2003

Breaking the German Dams: Flying into History, Newsdesk Communications Ltd, London, 2008

Websites

Avro Lancaster
History Channel
Imperial War Museum sound archive
Lancaster Museum, Canada
RAF Museum
World War II Experience

Videos/DVDs

The Dambusters, Windfall Films, Castle Home Video 2002
The Dambusters, Channel 4, Tigress Productions
The Dambusters, Daily Mirror DVD, World War II, Episode 6
The Dambusters Raid, Delta Music 2002

Index